FAITH EMBODIED

GLORIFYING GOD WITH OUR PHYSICAL *and* SPIRITUAL HEALTH

STEPHEN KO

ZONDERVAN REFLECTIVE

Faith Embodied

Published in Grand Rapids, Michigan, by Zondervan. Zondervan is a registered trademark of The Zondervan Corporation, L.L.C., a wholly owned subsidiary of HarperCollins Christian Publishing, Inc.

Requests for information should be addressed to customercare@harpercollins.com.

Zondervan titles may be purchased in bulk for educational, business, fundraising, or sales promotional use. For information, please email SpecialMarkets@Zondervan.com.

ISBN 978-0-310-15171-5 (audio)

Library of Congress Cataloging-in-Publication Data

Names: Ko, Stephen., author.
Title: Faith embodied : glorifying God with our physical and spiritual health / Stephen Ko.
Description: Grand Rapids, Michigan : Zondervan Reflective, [2024]
Identifiers: LCCN 2023042122 (print) | LCCN 2023042123 (ebook) | ISBN 9780310151692 (paperback) | ISBN 9780310151708 (ebook)
Subjects: LCSH: Health—Religious aspects—Christianity. | Well-being—Religious aspects—Christianity. | Christian life. | BISAC: RELIGION / Christian Living / Spiritual Growth | SCIENCE / Life Sciences / Human Anatomy & Physiology
Classification: LCC BT732 .K668 2024 (print) | LCC BT732 (ebook) | DDC 261.8/321—dc23/eng/20240119

LC record available at https://lccn.loc.gov/2023042122 LC ebook record available at https://lccn.loc.gov/2023042123
Published in association with Books & Such Literary Management, 52 Mission Circle, Suite 122, PMB 170, Santa Rosa, CA 95409-5370, www.booksandsuch.com.

Cover design: Gearbox Studio
Cover photos: © Shutterstock; Getty Images
Interior design: Sara Colley

Printed in the United States of America

24 25 26 27 28 LBC 5 4 3 2 1

Faith Embodied leads us to reflect on how the way God made us impacts our souls and spirits. It is an eye-opening look at things we often take for granted about the way we are made and function. Read it to learn much about who we are and how we can flourish.

–DARRELL BOCK, executive director for cultural engagement, Howard G. Hendricks Center for Christian Leadership and Cultural Engagement; senior research professor of New Testament studies, Dallas Theological Seminary

I've been in church leadership for more than two decades, and at the age of forty-one, I suffered a heart attack and had a quadruple bypass. I could blame it on stress and sedentary living, but ultimately, I could preach about healthy mind, body, and spirit better than I knew how to live it. Dr. Ko writes as a physician and a pastor and gives language to how to make the theology of health practical so that you can understand how much physical and spiritual healing are linked. Understanding this is integral to disciple-making in our time.

–DANIEL YANG, national director of churches of welcome, World Relief

Read Stephen Ko's new book, which links physical and spiritual wellness in a way you have never thought about before. You will look at things differently once you read his insights into both health and Scripture.

–JIM CYMBALA, senior pastor, The Brooklyn Tabernacle

This is a beautiful book filled with anecdotes drawn from Rev. Ko's unique combination of medical and pastoral experience. Rev. Ko fills the reader with appreciation for our bodiliness as part of God's purpose and incentivizes us to make the most of our senses and faculties by offering them in love and worship. A life-changing resource for individuals and churches.

–KIRSTEEN KIM, PhD, Paul E. Pierson Professor of World Christianity, and associate dean, Center for Missiological Research, Fuller Theological Seminary

As a medical doctor and a health practitioner, Dr. Stephen Ko acquaints us afresh with the glories of our human bodies. As a pastor and a theologian, he helps us marvel at how the *imago Dei* incarnated in each magnificent human body is wonderfully matched by the "exact image of the Godhead" manifested in the unique, incarnate Son of God. He reminds us that worship of the creator and sharing of his redemptive love are beautifully experienced and expressed, as God intended, in flesh.

–TIM CROUCH, vice president, Alliance Missions

Drawing on his rich background as pediatrician, epidemiologist, public health officer, professor, and pastor, Stephen Ko has made a significant contribution to a Christian understanding of the human body. *Faith Embodied* lucidly applies a biblical and theological understanding to spiritual development, worship, and everyday life. I highly recommend this readable, engaging book.

–DENNIS P. HOLLINGER, PhD, president emeritus and senior distinguished professor of Christian ethics, Gordon-Conwell Theological Seminary

It's not often one encounters a book that is a perfect blend of Scripture, theology, stories, and medical expertise. But Dr. Stephen Ko has provided readers with just such a book in *Faith Embodied*. You will learn so much as you read it, not just about the wonders of the physical body but also about the spiritual implications of what it means to be created in God's image. This is a deeply encouraging and insightful book with a holistic depiction of faith that will feed your mind, body, and soul.

–HELEN LEE, author and speaker

We need doctors for our bodies. We need them increasingly for our minds as well. With Stephen Ko, we have a doctor for body, mind, and spirit, someone who can gently probe various ways we are broken by sin and guide us to Jesus, who redeems. Read this book and discover more fully what it means to be fearfully and wonderfully made by God.

–DR. WALTER KIM, president, National Association of Evangelicals

As both pastor and medical expert, Stephen Ko is uniquely positioned to help us recognize, appreciate, and live in the way God made us. The God who designed, created, sometimes miraculously heals, and will someday raise our bodies cares about our physical health. Because he usually works through what he has already created, however, he invites us to be good stewards of the health resources he has already provided.

–CRAIG S. KEENER, F. M. and Ada Thompson Professor of Biblical Studies, Asbury Theological Seminary

After reading this book, you will be inspired to echo the gratitude and wonder of the psalmist: "Thank you for making me so wonderfully complex. Your workmanship is marvelous; how well I know it!" (Ps. 139:14 NLT). Dr. Stephen Ko, physician, epidemiologist, and theologian, writes as the Paul Brand of our day to integrate amazing insights into human anatomy, physiology, and pathology with biblical truth about our creator and the Christian faith. *Faith Embodied* will likely produce in readers a faith emboldened as they gain a greater understanding of how our bodies tell God's loving and redemptive story.

–DR. MICHAEL CHUPP, chief executive officer, Christian Medical and Dental Associations

Drawing upon his theological training and his medical expertise, Rev. Dr. Ko eloquently reflects on the integration of physical function with worship and Christian service. This book emphatically offers a remedy to the divorce of the physical and the spiritual.

–DANIEL K. ENG, assistant professor of New Testament, Western Seminary

In *Faith Embodied*, Dr. Ko leads us on a transformative journey to appreciate the way God can redeem us as whole human beings by becoming more aware of the beauty of the body's creation and the mystery of the Spirit's incarnation. He weaves a rich tapestry of evidence from science and biblical faith to inspire us toward the full-orbed embodied life that God intends us to live and proclaim as a global common need.

–DANIEL W. O'NEILL, MD, MTh, managing editor, *Christian Journal for Global Health*; assistant clinical professor, University of Connecticut School of Medicine

As a seminary dean and professor for twenty-five years, I have taught a course on divine healing many times. This book, *Faith Embodied*, by my dear friend Dr. Stephen Ko, is a gift from God. It brilliantly bridges the gap between the physical and the spiritual. I will be using it as a text in my next course.

—DR. RONALD WALBORN, former vice president and dean, Alliance Theological Seminary; executive director of urban initiatives, Asbury Seminary NYC

This book is an educational, informational, and motivating read. It has challenged me to look more closely at much of what I take for granted regarding our senses, creativity, rest, and worship. By grounding his descriptions and prescriptions solidly on the Word of God and his revelation in the incarnation of our Lord (looking to Jesus), Ko ensures this book's timeless value.

—RAJAN MATTHEWS, president, Alliance University

Faith Embodied offers a fresh call to love the Lord with heart, soul, mind, and strength. Dr. Ko combines a biological analysis of our five senses and movements with a call to a fully embodied life of worship. He invites us to embrace an integrated vision of health and spirituality flowing out of the divine intent for human beings created in the image of God.

—STANLEY JOHN, PhD, associate professor of intercultural studies (NYC), ESJ School of Mission and Ministry, Asbury Theological Seminary

In earlier times, pastors understood their work to be the cure of souls, in the same way that today's physicians understand their work to be the cure of bodies. Stephen Ko embraces both responsibilities. As a physician and a pastor, he delights in the intricate beauty and mystery our bodies and souls reveal. He discerns how our embodiment shapes our spiritual lives. He diagnoses where we ail. And he knows what *healthy* means. In *Faith Embodied*, Ko turns a doctor's eye and a pastor's heart to our lives—our gloriously and mysteriously embodied lives—and prescribes a way that leads to a life that glorifies God.

—GREG JAO, senior assistant to the president, InterVarsity Christian Fellowship/USA; author, *Your Mind's Mission* and *The Kingdom of God*

To Bridget, Topher, and Luke

For all the nights you slept alone,
For all the meals with an empty seat at the table,
For all the moments you experienced without me,
The sands of time can never be reclaimed,
But perhaps God will use this labor of love
To redeem the faith of others.

Love you.

CONTENTS

ACKNOWLEDGMENTS

Thank you, Cynthia Ruchti, for being a friend, mentor, editor, and agent extraordinaire.

INTRODUCTION

INCARNATIONAL HEALTH

Most of us treat our bodies as separate entities from our spirits, related in only minor ways—if at all. We worship on Sunday mornings, then gorge on an unhealthy lunch afterward. We watch our favorite Netflix guilty pleasure in the evenings, then switch to saying our nighttime prayers. If we are ambitious, we do our devotions and exercise in the morning, but rarely simultaneously. They're separate, unrelated activities, aren't they?

Have you ever tried listening to worship music while you exercise? Not the slow, heartfelt songs that touch your heart and bring you to tears at the end of a set but the upbeat melodies that inspire dancing and hand clapping. When I do this, it produces a synergistic effect. Running on the treadmill makes the lyrics come alive, as if I'm "raising a hallelujah" to the heavens. Hearing the perfect cadence of drumbeats among thoughtful words allows more emotional—and physical—weight to be lifted. When we live as if our physical and spiritual health are deeply and richly connected, both are strengthened.

Though we may separate the body and the spirit, ministry

and medicine have taught me the two are inextricably tied. Illness, disease, and accidents invariably affect our spiritual health. The reverse is also true. Spiritual health has a direct impact on physical well-being.

I am a scientist, a researcher, an academic, and a pastor. It is not often that a career path leads from pediatrician to epidemiologist, to public health officer for the CDC, to college professor, to senior pastor. That's an odd constellation of specialties, and you might wonder how I balance the different priorities of those disciplines. Can pastors trust medical research without compromising their faith in a healing God? Can scientists and physicians study the intricacies of the human body without losing an expectation for the miraculous and supernatural?

These questions are the point of this book. I believe my unusual career path was ordained by God so I can speak into the intersection of faith, medicine, and public health. My faith in Christ is enhanced, not diminished, by what I know about science and medicine. My medical and public health experiences are enriched by my foundational trust in the God of ages, the creator of all we know and have yet to discover about the human body.

Years of study and experience have taught me that the gap between physical and spiritual health is imaginary. One of my congregants suffers from chronic pain. No moment passes without sharp tingling sensations pulsating throughout her body. Though the suffering is constant, some days are worse than others. These are the darkest hours, when anger, bitterness, and resentment fill her heart. Nothing can be said to soothe her mind. In those moments, the anguish of her soul reflects the pain in her body. Just as my practice of adding worship music to exercise enhances both physical and spiritual health, so illness and disease can magnify spiritual malaise.

The interrelatedness of body and spirit is at the core of God's design for us—the concept of incarnation, a word we typically associate with Jesus' inhabiting human form. We don't worship a disembodied savior or a set of spiritual principles for right living or bettering our relationship with God. We worship Jesus, the eternal Son, who became human in every way we are. Jesus "took on flesh," we're told in the Bible (John 1:14). He inhabited every cell of human being, saw through human corneas, touched with fingers covered in human epidermal tissue, walked on feet with the same twenty-six bones, thirty-three joints, and about 120 tendons, ligaments, and muscles that yours and mine have. Jesus was incarnated in a human body for the highest and holiest of purposes, winning our hearts and paving the way for our salvation.

Then when Jesus returned to heaven, he sent his Spirit to inhabit us. When we acknowledge Jesus as our savior and lord, the Holy Spirit takes up residence in our whole selves—including our corneas, joints, tendons, and skin. We offer these things to him as we love, serve, and honor Jesus. In him we live and move and have our being, as the book of Acts tells us (17:28).

Through the power of the Holy Spirit, we are united to Christ. Just as the branches are part of the vine (John 15), we are linked to him. We join in his life, death, and resurrection. We live incarnationally[1] when the Holy Spirit acts in our unified bodies and spirits to worship and connect us to Jesus, the Incarnate One.

When we worship incarnationally, we allow our God-designed bodies—from our major organs and systems down to the finest microscopic details—to do what the Spirit came to do: guide, steer, remind, lead, fill, fulfill, love, teach, inspire, speak, and worship purely. We are as divinely infused by the Spirit of God as the babe in a manger, who grew up, served, sacrificed, hung on a cross, and rose to new life. When we make incarnational health decisions, we think about the ways our physical health influences our spiritual health, and vice versa. We stop pretending that the things we consume, the ways we use our bodies, and the things we do to take care of ourselves have nothing to do with our relationship with God.

What do we overlook if we fail to use our God-created eyes, ears, noses, mouths, and hands as instruments through which the Spirit can interact with and impact our world? What do we thwart if we fail to care for and protect our vision, our breathing, or our ligaments for the full time God allots for us here on earth? What do we miss if we assume an injury or illness evaporates God's purposes for us rather than finding new ways in the intricacies of God's marvelous creation to do what his Spirit instructs?

Not only are incarnational health, worship, and living possible, they are God's good design, and he is waiting for us to fully embrace them.

CORRUPTION OF GOD'S CREATION

But before we jump farther into the concept of incarnational health, we must look at its inverse. If we believe God designed our bodies carefully and called them good, then why do they seem to be constantly failing us? Why do they break down over time? Why are they susceptible to invasion by corruptive outside forces? What is the role of illness, disease, and aging in our lives?

Though we were designed to glorify God, sin often stands in the way. Since the fall, the wages of sin are death. Paul shared this reality in his letter to the Romans: "Sin came into the world through one man, and death through sin, and so death spread to all men because all sinned" (5:12).

From the dust we were formed, and to dust we will one day return. Choices made at the garden of Eden ensure it (Genesis 1). Across the globe, violence leads to murder, and wars bring death. Pandemics infect millions of people, and hundreds of thousands die. At the microscopic level, infectious diseases and chronic illnesses destroy our bodies. Let's look specifically at how cancer twists God's good design into something horrific and life threatening.

Normal cells grow, divide, and die when signaled to do so. Each cell plays an intricate and vital role within our organs and bodily systems. When functioning correctly, cells fulfill their roles without disturbing the work of other organs. In 1 Corinthians 12, Paul likens biological organs to roles within a church. Just as every organ is essential to a healthy body, each person is vital to a flourishing church. The foot, for example, cannot say to the hand that it does not belong to the body.

Yet that's what cancer does. It disregards the role of other cells. When cancer takes hold, cells no longer act in unity. Instead, they live autonomously with no regard to the healthy structures surrounding them. Cancer results in cells refusing to die when they are supposed to, dividing uncontrollably, spreading to surrounding tissues, and eventually forming malignant tumors.

Blood cells, skin cells, bone cells, muscle cells, and nerve cells were fashioned in unity. Each was created to live, function, and work in harmony with one another. Yet cancerous cells act in self-interest. They do whatever is beneficial for their growth and self-preservation. With time, they invade neighboring organs and destroy surrounding tissues. As you can see, there's an analogy in how these biological processes play out and how spiritual corruption can occur within the body of Christ.

How, then, should we understand the roles of illness, disease, and aging in God's created order? Protestant Reformer Martin Luther once said, "We would like to be rid of all our infirmity which, to our superficial conception, appear to be a great hindrance to doing useful things, and yet it is most questionable if we should bring forth any fruit unto God without them."[2] If Luther believed illness yields blessings, how do we explain his actions of healing during horrific plagues? In his opinion, disease was neither God designed nor God caused but permitted by God according to God's purposes.

In contemporary times, theologian Leonard Sweet views sickness and infirmity as a "power hostile to God that destroys life."[3] Yet Sweet believes suffering can sanctify our actions, cultivating compassion and empathy for others in this broadly evangelical perspective. For him, restoration and flourishing are not necessarily seen in perfect health and well-being but are found when humanity embraces its God-given roles—including healing. Sweet is careful not to "attribute the wreck of well-being to a single cause—divine command or satanic powers or punishment of sin." In doing so, we "simply allow people to avoid what nature, science, medicine, and religion offer for restoration."[4]

There is no easy explanation for the ravages of disease. We can be confident that illness, sin, and death were never part of God's plan for Adam and Eve. But because of their choices, consequences remain for all humanity. Nevertheless, like other forms of suffering, illness can foster perseverance, development of character, and hope in the Lord (Rom. 5:3–5). In a world where these awful things exist, we can take up God's call to live incarnationally by making good incarnational health decisions for ourselves and others.

TENSION WITHIN INCARNATIONAL HEALTH CHOICES

What do I mean by "incarnational health decisions"? Through illness, aging, pandemics, and natural disasters, we have countless opportunities to respond in faith, whether through the food and drink we consume, the images we see, or the voices we listen to. Each of our choices provides a chance to worship incarnationally, both individually and collectively.

These decisions provide an opportunity to glorify the Lord while standing against the corruption of sin. In Romans 11:36, Paul affirms the centrality of God in all creation, "for from him and through him and to him are all things." He is the source of our existence, the sustainer of our lives, and the goal of our work.

Incarnational health decisions present innate tensions. First is the idea of sanctification of the body—the process of being made holy. A natural tension exists between making choices that preserve our bodies as pure, healthy, and blameless and succumbing to the desires of our flesh. A second tension comes from the call to sacrifice our broken vessels for the sake of others. Will we surrender our bodies for the spiritual, mental, and physical health of others?

First Corinthians 6:19–20 teaches us that our bodies are temples of the Holy Spirit; therefore we should glorify God with them. Paul explains that these "jars of clay" are to be preserved as pure, undefiled temples of God. This includes fleeing from sexual immorality, the lust of the eyes, and temptations of the heart. When we commit adultery with our bodies, we deconstruct the natural relationships intended for our spouses. More important, we destroy the unity inherited from Jesus as members of Christ.

Yet not only do adultery of the body and perversion of the heart threaten the sanctifying work of the Holy Spirit. Conscious and unconscious decisions can also derail how we embody Christ. Seemingly innocuous images from social media can hijack our minds and steal our hearts.[5] Soon they become obsessions beholden to the neural circuitry within our brains. Instead of being witnesses for Jesus, we reflect whatever our eyes focus on.

That is the draw of sin and a tension in incarnational health choices. We feel a subtle urge to take another bite even though our bellies are straining against our belts. It's far too easy to drink more wine while our judgment is a little less than sensible. Some would say overindulgence is what makes us human: to eat, drink, and be merry is what gives meaning to our lives.

The earliest monastics understood these dangers all too well. They chose to live ascetically in the wilderness and later in monasteries. In radical acts of self-deprivation, monks and nuns embraced poverty, chastity, and obedience.

They left the trappings of the world to focus on communion with the Father in heaven.

But many Christians would argue these monastic choices were not incarnational, they were isolationist. Instead, we should sacrifice our lives for others so that they might experience Christ. This sacrifice presupposes the ministry of presence that Jesus modeled in life and death. When others distanced themselves from the sick and ill, he healed the blind and deaf. While religious laws recommended quarantine from the ceremonially unclean, Jesus reached out to touch lepers and bleeding women.

Following this model, Christians since the Cyprian and Antonine plagues of the first and second centuries have chosen to be vessels of healing for those in need.[6] During the bubonic plague, Martin Luther exemplified this sacrificial love. Instead of fleeing the Black Death in his hometown of Wittenberg, Luther and his family chose to stay and minister to the dying and sick. He reasoned that those at death's door needed a doctor and a shepherd to comfort their souls and administer sacraments and last rites.[7]

Centuries later, believers like Kent Brantly willingly risked exposure and death in treating highly infectious ebola patients. Brantly even returned to West Africa to treat more dying patients after he recovered from the disease himself.[8]

Why would he choose this path? Because the Good Shepherd lays down his life for his sheep, according to John. He calls each of us to lose our lives for his sake, to pick up the cross and follow him, and to define love by giving our lives for others. Each of us is invited to follow Jesus wherever he leads, even to Calvary. Sometimes incarnational health means taking steps to preserve our well-being. Other times it means setting our well-being aside for the sake of others.

CREATION IN THE IMAGE OF GOD

How do we reconcile the preservation of our bodies as holy and blameless with the call to sacrifice these earthen vessels for others? At the heart of this dilemma are the beauty of creation, the sanctity of life, and the call to glorify God.

All lives are sacred. Genesis teaches us that we are made in the *tselem* and *demuth* of God to worship him. These Hebrew words are usually translated into English as "image" and "likeness." From the Latin *imago Dei* comes "the image of God."

What does it mean when God says, "Let us make man in our image, after our likeness," as recorded in Genesis 1? According to John Piper, "Humankind was created to be a graphic image of the Creator—a formal, visible, and understandable representation of who God is and what He's really like."[9]

As a graphic image of the Creator, our bodies, minds, and souls should be transformed daily to reflect the Father. The fact that we represent God should influence how we see ourselves and others, even in a fallen world. No matter how severely our bodies are ravaged by disease or our relationships are marred by sin, we are created in God's image. We should view every organ, sensory adaptation, and musculoskeletal function with a sense of awe and wonder. Every person, relationship, and connection must be given respect and dignity.

We may never know the extent to which the human qualities used to describe God are literal or analogical. Still we can say with confidence that our capacities were created in his image. Heartful passions and desires allow us to love deeply. Beautiful minds consider the complexities of nature and science. Hands and feet are meant to engage in meaningful labor. Our organs provide a window to our creator, who is all powerful, everywhere present, and all knowing.

Healthy choices affirm the handiwork of God within our bodies. They engage different parts of our bodies (cells, tissues, muscles, bones) and different senses of our organs (sight, hearing, smell, taste). In doing so, they move us one step closer to the garden of Eden, where once we perfectly reflected the glory of God.

Our bodies' intricacy is astonishing, from invisible microscopic cells to large muscles that allow us to walk and run. This intricacy speaks to the inspiration of an awe-inspiring Creator, who formed our inward parts and knitted us together in our mothers' wombs (Ps. 139:13). King David writes that we are fearfully and wonderfully made (v. 14). We were created with a heartfelt interest in great reverence, unique and set apart for God's purposes.

As image-bearers, we were made to glorify God. In fasting, prayer, and meditation, we worship him. But running, walking, and swimming were also designed to give glory to the Lord. When we realize that God created exercise as a way to redeem our bodies for his purposes, we see the gym and pool in a different light. Instead of perfecting our bodies for the sake of others, we reflect the image of God so that others may worship him.

In contrast, sometimes we are called to sacrifice our bodies for others who are made similarly in his likeness. Donating blood, bone marrow, and bodily organs reflects the sacrifice of Christ. When we give our eyes, kidneys, hearts, and lungs, they become living witnesses of God's love for those who receive them. Often, these sacrifices of love result in flourishing for others.

Jesus encouraged the disciples to "heal the sick, raise the dead, cleanse those with leprosy, cast out demons. Freely you received, freely give" (Matt. 10:8 NASB). Organ, blood, and bone-marrow donations, as well as exposure to disease when treating others, are not only acts of charity but also profound ways to honor the gift of life God has given to us.

REFLECTION OF THE RESURRECTION

And as we pursue incarnational worship of God, what hope do we have when health is impossible in this life? Individual and collective healing, while good for their own sakes, are also reflections of our future resurrection. Only in the next lifetime will cancerous cells be reborn healthy and what was imperfect be made whole. While the resurrection of Christ gives meaning to our health choices today, the anticipation of our future resurrected bodies makes every health decision worthwhile, even if our circumstances look grim.

While working in the intensive care unit at Bugando Medical Center in Mwanza, Tanzania, I sensed death lurking everywhere. Third-world countries lack state-of-the-art diagnostics, costly ventilators, and lifesaving medications. Though I had never seen a case of malaria in the United States, cerebral malaria was common in the sickest wards of Africa.

Cerebral malaria is a severe neurological complication of *Plasmodium falciparum*, the deadly organism that causes malaria. Parasite-infected red blood cells accumulate within the brain, and after a few days of fever, children experience seizures and end up in a coma. The prognosis is often grave because of brain swelling and increased blood pressure in the skull.[10]

Without treatment, cerebral malaria is fatal. But even with antimalarial drugs, up to 25 percent of children still die. Many of those who survive live with severe brain damage, blindness, difficulty moving, long-term behavioral issues, and epilepsy.

What hope could an American doctor offer to Tanzanian mothers whose babies were diagnosed with cerebral malaria? The medical prognosis for their young children was awful. Their only hope was that of the future glory described in Revelation. On that day, God "will wipe away every tear from their eyes, and death shall be no more, neither shall there be mourning, nor crying, nor pain anymore, for the former things have passed away" (Rev. 21:4).

These words of John often brought tears to the eyes of young mothers. Believers knew in their hearts it was true, but it was still difficult to imagine the death or disability of their once vigorous children. In tremendous pain, their only comfort was knowing that one day they would see their babies with unbroken bodies again—strong, vibrant, and thriving. In those moments, the words of Paul rang true: "O death, where is your victory? O death, where is your sting?" (1 Cor. 15:55).

Though humanity is destined for death, one day we will inherit resurrected bodies immune to cancer, disease, plague, and the effects of aging. *Anastasis*, the Greek word for "resurrection," also means "to stand again." Just as Jesus Christ rose, so too will our bodies arise again.

Paul describes this miraculous day of awakening: "So is it with the resurrection of the dead. What is sown is perishable; what is raised is imperishable. It is sown in dishonor; it is raised in glory. It is sown in weakness; it is raised in power. It is sown a natural body; it is raised a spiritual body. If there is a natural body, there is also a spiritual body" (1 Cor. 15:42–44).

The process of aging begins early in our lives. After age twenty-five, more cells perish in our bodies than are created. Slowly, our bodies lose control of their DNA, the genetic blueprint of individual existence. We become more prone to cancer, autoimmune disease, and other illnesses. Nerves, tissues, and organs waste away insidiously.

But good news awaits: our heavenly vessels will be glorious! Resurrection bodies will be imperishable, immune to disease, illness, and aging. We will live impervious to the corruption of sin. Jesus' transfiguration foreshadowed this future glory. At that time, Peter, James, and John witnessed the radiance and wonder of the Son of Man. A similar glory will be reflected by our resurrected bodies.

On that day, they will be raised in power. No longer will our bodies suffer developmental disabilities or debilitating diseases. Our muscles and bones will not succumb to the challenges of the world. Lou Gehrig's disease will not erode our nervous systems and spinal cords. Parkinson's will not steal our ability to move.

In the fullness of time, we will be sanctified wholly. Our souls will be in harmony with the Holy Spirit. No longer will we be subjected to the weakness of the flesh. Instead, our resurrected bodies will work in concert with the spiritual desire to worship God.

WORSHIP THROUGH INCARNATIONAL HEALTH CHOICES

In this lifetime, though, we all need healing—personally, collectively, and globally. When we mourn, God meets us in our grief. In the desolation of isolation, Jesus heals our brokenness.

Just as he had compassion for the crowds, Jesus has mercy for us. He healed every disease and affliction during his earthly ministry, and he still desires to heal every sickness and infirmity affecting us today. He alone was pierced for our transgressions, crushed for our iniquities; by his wounds, we are healed (Isa. 53:5).

The incarnate Christ spent most of his time with sinners, tax collectors, and the ill. He healed the blind, lame, mute, bleeding, and demon possessed. Why did he travel throughout the towns healing every sickness and disease? Jesus proclaims in Luke 5:31–32, "Those who are well have no need of a physician, but those who are sick. I have not come to call the righteous but sinners to repentance."

The miracle of healing reminds us what we look forward to in future glory. For some, God grants the firstfruits of healing within our present lives. They are signs of the harvest to come with the resurrection of our bodies, but not necessarily a promise we will all inherit while alive.

Let's look at one example of an incarnational health choice. In an age of gluttony and overindulgence, the spiritual discipline of fasting is countercultural and neglected. When every ad on TV or our phones encourages us to seek comfort and contentment at all costs, the idea of depriving ourselves seems radical. Without food and water, the human body can survive only a week or two. Soon, the person fasting feels tired and experiences headaches and symptoms of malnutrition. Why would anyone choose to experience these things? Yet for forty days and forty nights, Moses fasted while God gave him the Ten Commandments. Exodus 34:28 informs us that he neither ate bread nor drank water during this time. Instead, fasting drew him closer into intimacy with God.

Fasting is astonishing at a macroscopic and microscopic level. When fasting, the body has limited access to sugars through regular intake. Instead, cells are forced to extract glucose from within and produce energy independently. During gluconeogenesis, the body converts amino acids, lactate, and fat into essential sugars. In addition, the body begins to burn off stored fat in a process called ketosis.

With patience, practice, and time, those in ketosis no longer feel shaky, weak, or lightheaded. Soon, energy levels remain steady and consistent throughout the day. A decrease in overall hunger results in increased mental focus, physical energy, and spiritual acumen. Not only does fasting cleanse our stomachs, but it also purifies our souls. Choosing to deprive ourselves of something our bodies crave leads us into greater conformity with God's image.

Other incarnational health choices lead to similar spiritual outcomes. Prayer focuses our eyes on the kingdom while limiting our view of the trappings of the world. Meditative breathing extinguishes stress and anxiety while fostering communion with God.

But these health decisions should not be made with only the individual in view. They provide communal healing as well. When we cooperate with God as instruments of his healing, we affirm Jesus' desire to heal the broken. Not only do we embrace the prayer of faith found in James 5:16 but we also use our God-given knowledge, intelligence, and wisdom to discern how to protect the sanctity of life. Our communities benefit when we learn to see the intertwined physical and spiritual dimensions of health.

By embracing incarnational health in our lives and the lives of others, we worship God while pointing others to the life, death, and resurrection of Jesus. Ultimately all our bodily decisions should bring glory to God. For "whether you eat or drink, or whatever you do, do all to the glory of God" (1 Cor. 10:31).

Incarnational health choices allow others to appreciate the majesty of the Creator. They help us catch a glimpse of the *imago Dei* before the corruption of sin metastasizes throughout our bodies. Whenever we cleanse ourselves of what is dishonorable, we will be vessels for honorable use, set apart as holy (2 Tim. 2:21). Others will see beyond temptations of the flesh, realizing instead the potential to be temples of the Holy Spirit.

In the pages that follow, we will explore the myriad ways in which God created our physical and spiritual health to work in tandem. We will see how he designed our five primary senses to help us worship him, what we lose when those senses don't work as designed, and the spiritual implications of an incarnational appreciation for our eyes, ears, noses, nerves, and taste buds. Then we will look at the extraordinary ways in which God designed us to navigate and engage our environments and the people in them. With a better understanding of how human beings breathe, move, create, love, and rest, we will capture God's vision for our purpose in the world and the ways we can stay healthy—physically and spiritually—to better offer health to others.

As we learn about the biological processes that help us experience and interact with the world around us and the medical implications when they don't work as intended, we will get a better sense for how those things help us relate to God and how neglecting their healthy functioning can lead to spiritual corrosion as well.

Faith, science, and medicine are not incompatible. Instead, they corroborate absolute truths seen in nature, human cognition to research and explain, and the ethical mandate to sacrifice for the oppressed while caring for our neighbors. By increasing our awareness of the world around us, including the biological processes within our bodies, we can make better choices that positively impact the world around us.

God heals through supernatural power and natural laws that govern science and medicine. Through the incarnation of his Son, he chose to save humanity. By the indwelling of the Holy Spirit in those made in his image, he elected to improve individual health and collective well-being. The choice before us is whether we will cooperate with the Spirit as we grow in the Son's likeness by taking steps to improve our incarnational health and care for the incarnational health of those around us.

NOTES

1. Throughout this book, when I use the term *incarnational*, I do not mean the union of divinity and humanity in Jesus Christ. Instead, I'm referring to the embodiment of how Jesus would act, live, and speak. This is made possible by the Holy Spirit dwelling inside us. Incarnational health decisions glorify God with our bodies, which are temples of the Holy Spirit. This analogy is not meant to equate the Spirit's indwelling with Jesus' incarnation.

2. Luther, M. (1979). *The table talk of Martin Luther.* Baker.
3. Sweet, L. I. (1994). *Health and medicine in the evangelical tradition.* Trinity.
4. Sweet, L. I. (1994). *Health and medicine in the evangelical tradition.* Trinity.
5. Carson, V., & Janssen, I. (2012). "Neighborhood disorder and screen time among 10–16 year old Canadian youth: A cross-sectional study." *Int J Behav Nutr Phys Act 9*, 66; Madigan, S. et al. (2019). "Association between screen time and children's performance on a developmental screening test." *JAMA Pediatr 173*(3), 244–50.
6. Bruun, C. (2007). The Antonine plague and the "third-century crisis." In Hekster, O., de Kleijn, G., & Slootjes, D. (eds.). (2007). *Crises and the Roman Empire: Proceedings of the Seventh Workshop of the International Network Impact of Empire (Nijmegen, June 20–24, 2006).* Brill; Harper, K. (2017). *The fate of Rome: Climate, disease, and the end of an empire.* Princeton University.
7. Luther, M. (1999). *Luther's works: Vol. 43. Devotional writings II.* Fortress.
8. Tryggestad, E. (2014, August 21). The inside story of ebola patient Dr. Kent Brantly's decision to serve in Liberia. *Christian Chronicle.* https://christianchronicle.org/dr-kent-brantly-is-who-he-is-because-he-tries-to-live-according-to-gods-will
9. Piper, J. (1971, March). The image of God: An approach from biblical and systematic theology. *Studia Biblica et Theologica.*
10. Idro, R. et al. (2010). Cerebral malaria: Mechanisms of brain injury and strategies for improved neurocognitive outcome. *Pediatr Res 68*(4), 267–74.

PART 1

THE GIFT OF SENSE

Unless we are ill or were born without one of the gifts of sense, it's easy to take our senses for granted. Masterfully designed and exquisite in function, each sense serves a purpose. They allow us to perceive the world around us. They protect us from harm. But our eyes, ears, mouths, noses, skin, and other sensory organs were created for so much more. They invite us to marvel at the majesty of creation, the smell of fresh roses, the taste of fine wine, the sight of clear blue waters, the sound of symphonies, and the sweet touch of a newborn babe. Through each, we can glorify God in extraordinary ways.

In part 1 of this book, we will consider the five basic human senses: sight, hearing, taste, smell, and touch. The organs associated with these senses send information to the brain allowing individuals to understand and interpret the surrounding world. The physiology that underscores each sense reflects the complexity of its design. Each chapter will teach us just how difficult it is to function without our senses. Not only will we learn how each sense works, but we will strive to understand its impact on our daily lives.

The senses aren't just for interacting with the world but play a vital role in each person's emotional and spiritual well-being as well. Touch conveys compassion. Smells evoke vivid memories. For better or for worse, what we experience through our senses can transform our hearts. By exploring the spiritual significance of each sense, we will understand how God designed these physical traits to help us worship and glorify him.

1

GAZE

The soccer ball ricocheted off another player, landing directly in front of my twelve-year-old niece, Madeleine. In an instant, she raced ahead from her midfield position, scanning the field for an open teammate. Then, out of the corner of her eye, she caught the figure of a fellow player streaking toward the opposite goalpost.

Anticipating the path of the fast-moving striker, she visualized the needed trajectory of the ball. Mustering all her strength, she kicked the black-and-white sphere just before a defender arrived to snatch it away from her.

The high-arcing ball landed perfectly in front of her teammate. With no defenders to beat, she scored in the left corner of the goal—just beyond the outstretched hands of the opposing goalie.

From a young age, Madeleine loved playing soccer. The camaraderie of team sports appealed to her. When given the option of what position to play, she chose midfield. It was the ideal position for setting up her teammates for success. But she was also prolific at scoring goals, once celebrating an improbable hat trick during a hard-fought game.

For Madeleine, school came easy. Classwork was a breeze. No matter the subject, she excelled. By the age of five, she was fluent in three languages. She had no greater joy than scanning the bookshelves at Barnes and Noble and finding a hidden gem to peruse. Her future seemed bright. The sky was the limit.

But just before her thirteenth birthday, her eyesight deteriorated. Both of her parents suffered from nearsightedness, so the family expected that Madeleine would inherit myopia. Yet when her visual fields started diminishing—like curtains closing from the sides of the stage before her eyes—her parents knew something was dreadfully wrong.

Only a handful of pediatric neuro-ophthalmologists exist worldwide. Her family located one at Children's Hospital of Philadelphia. Without an underlying cause of glaucoma or an impinging tumor, Madeleine's condition was diagnosed by the specialist as optic nerve atrophy.

The optic nerve gives life to our vision. A vast network of nerve fibers, it serves as a communication cable between our eyes and brain. Damage to the optic nerve typically results in permanent loss of vision.

Unfortunately, there is no treatment. Blindness is often inevitable.

Madeline's only option? To wait and see.

THE MIRACLE OF EYESIGHT

Too often, we take our vision for granted. Yet the eyes are one of our most precious organs.

Each eye is composed of more than one million optic nerve cells and more than one hundred million photoreceptor cells. Together they provide more information about our surroundings in a few seconds than all our other senses combined in several minutes. A picture really is worth a thousand words—or scents, tastes, and textures. Our eyes help us make split-second judgments while we walk, run, and drive. They also provide innumerable visual cues—nonverbal forms of communication—from the people we interact with. Up to 65 percent of all interactions come from body language.[1] That's why counselors are sensitive to the physical expressions of those they treat. Body language provides nuance to our communication that cannot be derived from words alone.

Sometimes facial expressions reveal our true feelings even when spoken words tell a different story. If a person looks directly into your eyes, they're in tune with the words you say. But breaking eye contact may mean they're distracted, embarrassed, or even ashamed. Crossed arms and a closed posture indicate hostility, anxiety, or unfriendliness. Open postures reveal friendliness and a willingness to work together.

How do our eyes process a friend's facial expressions and translate the complex movements of athletes we admire?

Images are made of light. When our eyes open, rays of light bounce off objects and people in front of us. Our eyes capture these wavelengths as they hit the cornea, the clear protective outer layer of the iris.

Then the light continues through the iris and pupil. The iris controls the quantity of light that enters the eye by opening and closing the black entrance—the pupil. After penetrating the pupil, light moves directly through the lens, which flips the image upside down and focuses it on the back surface of the eye, the retina. The shape of the lens adjusts for objects that are closer or farther away, allowing facial expressions and words to come sharply into focus.

When light arrives at the retina, millions of photoreceptor cells transform it into electric impulses. These messages are carried to the brain via the optic nerve. As they arrive, light is translated into the images we perceive flipped right side up.

Think of the camera on your smartphone. The process of translating light into information is similar. Yet the eye is far more complex, detecting and deciphering vast amounts of information every minute, hour, and day. A full 50 percent of the brain's function is dedicated to sight.[2] Vision is an extraordinary gift from the maker of all creation.

The story of the blind men in Matthew 9:27–31 shows just how meaningful vision is to those without it. By touching their eyes, Jesus heals two blind men and cautions them to keep the improbable news to themselves. Imagine their joy when they see the facial expressions of family, view the food they eat, and differentiate hues of color—all for the very first time. Who could keep that quiet?

When children learn they are nearsighted, they often lament the need to wear glasses. But after being fitted with their new lenses for the first time, little boys and girls are amazed by their newfound eyesight. Dull, shadowy figures suddenly become distinct and clear. Letters on the whiteboard are no longer fuzzy but come into focus. A once blurry world is full of clarity and radiance.

When Jesus heals the blind men, their lives of begging become full of possibility. With these new prospects, it's easy to understand why they can barely stay silent. They've just won the lottery of healing. So naturally, they shout for joy at the top of their lungs.

THE WINDOW TO CREATION

Our vision provides clues about the environment around us, and the beauty of nature captivates our souls. Though some do not yet know God, we can perceive his eternal power and divine nature in creation (Rom. 1:20). Through this general revelation, all of humanity can experience the glory of God. It is the Lord's way of revealing himself to each of us.

According to the psalmist, "The heavens declare the glory of God, and the sky above proclaims his handiwork" (Ps. 19:1). At the dawn of creation, God created the heavens and the earth. Then he separated the sky, land, and sea. Majestic canyons, pristine beaches, and colorful rain forests were born.

As he gathered dry ground, the land produced seed-bearing plants and trees—for food, for shade and protection, for medicinal purposes, and purely for beauty. Today, breathtaking flowers continue to capture our imagination. We earmark colorful bouquets for special occasions. Hydrangeas, orchids, spikenard, carnations, and magnolias bring floral art to life. Every year, countless people travel hundreds of miles to witness cherry trees blooming in spring and aspen leaves turning gold in fall.

God also made the waters swarm with living creatures and birds to fly high above the earth across the expanse of the heavens (Gen. 1:20). During a healthcare missions trip in Mwanza, Tanzania, I participated in a seven-day safari across Serengeti National Park. Though high-quality photographs and ultra-high-definition televisions provide spectacular images of animals, such images pale in comparison with seeing wildlife in person.

Serengeti National Park teems with life. More than

fifteen thousand square kilometers in size, it houses more than ten million hoofed mammals, including the "big five": lions, rhinoceroses, leopards, elephants, and Cape buffalos. During safaris these creatures come within inches of four-wheel-drive vehicles. Seeing these ferocious mammals in the wild, adjacent to giraffes and hippopotamuses, is awe inspiring.

In the Serengeti, vast expanses of highland plains and savanna woodlands allow wildlife to coexist with the Maasai people. These seminomadic sheep-herders are among the most famous people groups globally because of their proximity to game parks, their vibrant clothing, and their unique customs. They continue to practice livestock grazing as they live and migrate in northern Tanzania.

People who become up-close eyewitnesses to the Maasai way of life—to the wonders of their people and their culture, great and small, word-spoken by God or handcrafted by the people he created—gain new insight into God's image in humanity. To behold the splendor of creation with our eyes is an act of worship.

"The eyes are windows to the soul." This common phrase underscores the connectedness of our eyes to the whole self. On the surface, the eyes mirror our emotions and thoughts. Yet what we see has a profound impact on our spiritual health as well.

King David asked God to open his eyes that he might behold the wondrous works of the law (Ps. 119:18). When we refrain from gazing on idols that corrupt our bodies, we purify our hearts to worship God. When we open our eyes to the Word of the Lord, we worship in spirit and truth. Blessed are we when our hearts are pure, Jesus said, for we will see God (Matt. 5:8).

During the Second Great Awakening, camp meetings led to widespread revival among Protestants. These outdoor worship services were held in large, open areas under the shade of majestic oak trees. Over time, open-air tabernacles were established. Often, they were near scenic bodies of water in idyllic settings. Families camped near these sites for the duration of the revival meetings. Worship services were held from sunrise to sunset. Itinerant speakers preached the Word of God nonstop, often late into the evening.

These revivals had a tremendous impact on the spiritual lives of believers while helping to bring nonbelievers to faith. Congregants listened to the Word of God spoken by powerful preachers in a setting that allowed them to see the splendor of creation all around. Just like the Israelites in Nehemiah, they dwelled in the temple of God while gazing on the beauty of the Lord.

That's the essence of Paul's teaching in Acts 26:18: to open our eyes that we may turn from darkness to light and from the power of Satan to God. It is the beginning of revival.

Psalm 19:1–4 demonstrates how the heavens declare the glory of God and the sky proclaims his handiwork. Removed from everyday temptations of the world, believers were able to worship God surrounded by his revelation in nature.

It is a reminder to us that worship is not confined to church buildings and sanctuaries. Every moment we are outside is an opportunity to acknowledge and praise the creator of the heavens and the earth. Every time we open our eyes, we witness the majesty of creation. With each blink and purposeful gaze, the Holy Spirit can stir revival within our hearts. As we glorify God through our vision and encourage others to do so, they too can experience Jesus' presence.

THE LAMP OF THE BODY

According to Jesus, the eye is the lamp of the body. If your eye is healthy, your whole body will be full of light (Matt. 6:22). In a literal sense, the eye is the window through which light enters the body. Most people's awareness of objects, people, and God comes largely from the eyes. Without healthy vision, the body is devoid of light.

Globally, more than 2.2 billion people have a vision impairment.[3] Ninety-three million people in the United States are at high risk for severe vision loss, yet only half visited an eye doctor in the last year.[4] Ophthalmologic diseases impose an enormous financial burden—and that's not including loss of productivity. Uncorrected nearsightedness alone accounts for an estimated $244 billion worldwide. In East Asian countries, almost 90 percent of young adults are nearsighted. And across the globe, rates of myopia in the last several decades have increased at an astounding rate.

I can't remember a time when I had perfect vision. I have worn glasses since elementary school. During those years, I grew accustomed to heavy Coke-bottle lenses slipping off my nose. Today, my nearsightedness is so dreadful that I can read a book only if it is a few inches in front of my face. A centimeter farther away and everything becomes blurry!

What's the culprit for the increasing prevalence of myopia? For many, it's what we spend countless hours engaged with—looking at computer screens and smartphones. Screen time has long been linked to the epidemic of nearsightedness.[5] With a shift toward the urbanization of society, greater emphasis has been placed on those activities with less time spent outdoors.

The nearly universal adoption of digital screens among

children means that society is even more dependent on close visual activities at a young age. As digital natives, the average member of Generation Z in America spends more than eight hours a day on-screen.[6] Nearsightedness is now diagnosed at a much earlier age of onset and accompanied by accelerated progression in severity. As a result, the future will see higher burdens of severe myopia and an increased incidence of blindness.

How does screen time result in myopia or worsening nearsightedness? Think about how close you typically hold your smartphone in front of your eyes. Physical books are often read at an arm's length away, but smartphones inch nearer and nearer to our eyes to compensate for small screens and differing font sizes.

These factors place significant pressure on the early childhood development of what scholars call accommodation, the eye's focusing mechanism, and vergence. Vergence is the simultaneous movement of both eyes in opposite directions to maintain binocular focus, and that undue pressure can result in axial length elongation.[7] Eye length is the primary determinant of nearsightedness not caused by a genetic syndrome because it represents the thickness of the lens. And as eyes lengthen, the severity of nearsightedness increases.

In addition, screen time typically occurs indoors, where there is much less ambient light. This leads to less sunlight-induced retinal neurotransmission, which can impede the natural growth of eyes in developing children.[8]

There is a direct correlation between increased screen time and nearsightedness. Among five-to-fifteen-year-olds in India, the risk of myopia was eight times higher with greater than two hours per day of screen time compared with less than two hours per day.[9] Similarly, nearsightedness was markedly more common in Irish children who used their smartphones for more than three hours per day compared with less than one hour per day.[10]

Over the last four decades, the digital age has fundamentally changed how we see the world.

THE THEFT OF THE SOUL

Though screen time has harmful effects on our visual acuity, the more troubling issue is the way it hijacks our minds and steals our hearts. Research shows that social media is just as addictive as gambling.[11] Up to 10 percent of Americans meet criteria for social-media addiction, characterized by uncontrollable urges to log on or follow others. Doing so leads to excessive amounts of time online, to the detriment of other daily activities.

Innocuous pictures on Instagram become obsessions. Facebook posts cause stress and worry. Fear of missing out (FOMO) results in constant checking of Twitter and Instagram feeds.

Before long, we are immersed in a world that's hard to leave. Not everyone is addicted. Some wield discipline and practice self-control. But to a degree, all images engage our hearts. The more we allow them to dominate our field of vision, the less we see God.

Jesus cautions against exposing our eyes to darkness. He says that by looking at others with lustful intentions, we have already committed adultery in our hearts (Matt. 5:28). So likewise, by coveting the possessions of others, we store treasures only on earth where moths and rust destroy.

Pornographic imagery changes the habits of our minds while defiling our hearts. It distorts the perception of affection and love while objectifying others and denigrating their worth. When we view pornography, chemicals released in the pleasure-forming centers of the brain lead to compulsion over time. Emotional detachment accompanies feelings of desensitization, boredom, and even aggression.[12]

That's the subtle danger of sight. What we see can transform our hearts, for better or worse. Images are easily hardwired in our minds and etched on our hearts. Sight can easily lead to metaphoric gazing. What we see grabs our attention, often called "spotlighting." Our vision facilitates complex nonvisual needs for affirmation and belonging. The psalmist warns us not to set before our eyes anything that is worthless (Ps. 101:3).

In the Beatitudes, Jesus says, "Blessed are the pure in heart, for they shall see God" (Matt. 5:8). Purity of heart is core to the Christian faith. Words are heard and actions are witnessed, but hidden feelings and secret thoughts reside in our hearts. That's why, while humanity too often views and puts greater stock in outward appearances, God looks at our hearts (1 Sam. 16:7).

Online platforms, social media, and imagery can steal away our hearts. They do so by engaging us in seemingly innocuous ways. What we consume replicates the neural circuitry of other addictions. When we see that we have received new likes, shares, and retweets, our brains trigger the release of dopamine, similar to recreational drugs. Unbeknownst to us, gazing at cleverly designed pages and icons can marshal all of our attention toward things that might not be deserving of it.

Our eyes were created to witness the glory of the Lord.

When we cannot see Yahweh—when our eyes stare too often at lesser things—we cease to worship him. The greatest desire of the psalmist is to dwell in the house of the Lord and gaze upon his beauty (Ps. 27:4). Should that not be our overarching desire too?

THE PRESERVATION OF SIGHT

How do we keep our eyes healthy to appreciate all of creation? Though our bodies are temples of the living God, some parts of our bodies are more prone to destruction than others. Unfortunately, the eyes are among the most fragile.

First, we can limit our screen time. Two-thirds of practicing pediatric ophthalmologists report prescribing less screen time to control myopia.[13] Doing so not only preserves our vision but also guards our hearts against temptation.

Second, although regular exposure to sunlight cues specific areas in the retina to produce serotonin, lifting our moods while decreasing the risk for depression and anxiety, unfiltered ultraviolet rays from the sun can damage our eyes. Sunglasses that block 100 percent of UV rays can circumvent sun damage while allowing us to enjoy the natural revelation of God.

Third, nutrition choices that promote ophthalmologic health can stave off the effects of age-related vision issues such as cataracts and macular degeneration. These include green leafy vegetables; fish containing omega-3 fatty acids; nonmeat protein sources such as eggs, nuts, and beans; and citrus fruit such as oranges. Healthy diets also diminish obesity and the resultant risk for diabetes, which is the number-one cause of blindness in adults.

Annual visits to the optometrist help diagnose eye problems before they become permanent issues. If a doctor can catch retinal detachment, diabetic retinopathy, or corneal diseases early, they can be treated effectively.

Yet even for those with perfect 20/20 vision, our sight is limited. The visible light spectrum is only a sliver of the electromagnetic spectrum, between 380 and 700 nanometers in wavelength. Other portions have wavelengths too small or too large for our eyes to capture.

And the day will come when even the wavelengths we can naturally see will grow dim. For some, blindness comes like a thief in the night. Others' eyesight erodes gradually over a lifetime.

What will we do when we cannot see? How will we react when our lives are out of focus?

Paul encourages us to walk by faith and not by sight (2 Cor. 5:7). In the face of uncertainty, hopelessness, and despair, we look to Christ. The more we see him, the more we flourish in faith. For it is by the assurance of things hoped for and the conviction of things not seen that we hold on to hope (Heb. 11:1).

Henri Nouwen once wrote, "By gazing at Jesus, walking on earth, we give him living attention, and we see with our minds and hearts how he is the way to the Father."[14] These words echo those of the author of Hebrews—that we should "[look] to Jesus, the founder and perfecter of our faith, who for the joy that was set before him endured the cross" (Heb. 12:2).

Our spiritual eyes are just as important as our physical ones. Yet we are more prone to walk by sight than by faith. The way we react to our circumstances reflects how we see.

When we view the world with only our biological eyes,

we are susceptible to harmful emotions and wayward thinking. Yet our spiritual eyes yield peace in Christ, courage in hope, and trust in the sovereignty of God.

Reading the words of the gospel helps us focus on the life of Jesus. Gazing on the cross reminds us of his sacrifice for humanity. By remembering the life of Christ, we focus on the kingdom of God.

When the eyes of our hearts are enlightened, we experience the hope to which we are called (Eph. 1:18).

THE NEEDS OF OTHERS

When we fulfill the will of God, we worship him. In Matthew 25:35–36, Jesus said that the King will say to the righteous, "I was hungry and you gave me food, I was thirsty and you gave me drink, I was a stranger and you welcomed me, I was naked and you clothed me, I was sick and you visited me, I was in prison and you came to me."

Seeing the needs of others transforms us from selfish to selfless. It brings us from a place of pride to humility. The desires of the world become less enticing as we spend more time witnessing the poverty of others. We are convicted to act or driven to ignore.

When we see the world through Jesus' eyes, we invite the Holy Spirit to empower our lives. In time, we become like the Son of Man, who came not to be served but to serve others.

When Madeleine's vision began deteriorating, her mother took her on a hike in the Upper Peninsula of Michigan. As they were hiking, Madeleine's mother had to find a long makeshift walking stick for Madeleine and guide her on some of the more challenging parts of the trail.

At the top of the mountain, Madeleine said, "Wow, Mama. It's so beautiful up here. I mean, I can't see it all that well, but I can tell there's a lake down there, and there is beauty all around me."

They had just received news that Madeleine may eventually lose her sight completely, so they talked about the things she might want to do and the places she might want to go before that happened. She described wanting to go to Banff in Alberta, Canada, to hike and see the scenery.

Another part of her wanted to help people as much as she could while she was able: tagging along with her mom to Francophone Africa—where her mother had worked in the past—to translate for American doctors. She wondered whether she could bring a guide dog with her, and they laughed about which kind might cause the villagers the least fear.

Though Madeleine's eyesight was poor, her focus wasn't merely on her circumstances. She may not have had good physical vision, but she had inherited God's heart—a vision to love others.

For those experiencing horrific suffering, it's easy to be consumed by anger, bitterness, and resentment. As a result, even Christians feel justified to curse God at times, as the wife of Job recommended (Job 2:9).

Madeleine and her family chose to trust in the sovereignty of God. Though she could not fully appreciate the beauty of God's creation, Madeleine considered the needs of others. Their poverty was more significant than her desire to see great sights.

Poor eyesight became a blessing instead of a curse. Suffering led to perseverance instead of hopelessness. An inability to focus on the things of the world resulted in a greater capacity to gaze upon Jesus.

When we sacrifice our lives for the sake of others, we see Jesus. In time, everyone will see Christ when he reappears in glory (1 John 3:2). Yet those who fulfill his will see no longer as in a mirror dimly but as if they are with him face to face (1 Cor. 13:12).

So let your eyes look forward directly, and your gaze be straight before you, that you may see him clearly (Prov. 4:25).

QUESTIONS FOR REFLECTION

1. In what ways can you worship God with your gaze?
2. What subtle and overt temptations does your eyesight present to your heart and mind each and every day?
3. Matthew 6:22 says, "The eye is the lamp of the body." What does this mean, and in what other ways does the Bible speak about vision?
4. How can we redeem our physical and spiritual eyes for the glory of the Lord?
5. What specific needs do you see in the lives of friends, family, and nonbelievers? How is the Holy Spirit convicting you to engage?

NOTES

1. Pease, A., & Pease, B. (2006). *The definitive book of body language*. Bantam.
2. Sells, S. B., Fixot, R. S. (1957). Evaluation of research on

effects of visual training on visual functions. *American Journal of Ophthalmology.*

3. World Health Organization. (2021). Blindness and vision impairment. https://www.who.int/news-room/fact-sheets/detail/blindness-and-visual-impairment
4. Centers for Disease Control and Prevention. (n.d.). Common eye disorders and diseases. Retrieved 2021, from https://www.cdc.gov/visionhealth/basics/ced/index.html
5. Dirani, M. et al. (2019). From reading books to increased smart device screen time. *Br J Ophthalmol 103*(1), 1–2.
6. Suciu, P. (2021, June 24). Americans spent on average more than thirteen hundred hours on social media last year. *Forbes.*
7. Bababekova, Y. et al. (2011). Font size and viewing distance of handheld smart phones. *Optom Vis Sci 88*(7), 795–97; McCrann, S. L., Loughman, J., Butler, J. S., Paudel, N., & Flitcroft, D. I. (2021, January). Smartphone use as a possible risk factor for myopia. *Clin Exp Optom 104*(1), 35–41. doi: 10.1111/cxo.13092
8. Zhou, X. et al. (2017). Dopamine signaling and myopia development: What are the key challenges? *Prog Retin Eye Res 61*, 60–71.
9. Singh, N. K. et al. (2019). Prevalence of myopia and associated risk factors in schoolchildren in north India. *Optom Vis Sci 96*(3), 200–205.
10. McCrann, S. L., Loughman, J., Butler, J. S., Paudel, N., & Flitcroft, D. I. (2021, January). Smartphone use as a possible risk factor for myopia. *Clin Exp Optom 104*(1), 35–41. doi: 10.1111/cxo.13092
11. Hilliard, J. (2023, April 3). Social media addiction. Addiction Center. https://www.addictioncenter.com/drugs/social-media-addiction/

12. National Institute on Drug Abuse. (2022, March 22). Drugs and the brain. https://nida.nih.gov/publications/drugs-brains-behavior-science-addiction/drugs-brain
13. Zloto, O., Wygnanski-Jaffe, T., Farzavandi, S. K., Gomez-de-Liano, R., Sprunger, D. T., & Mezer, E. (2018). Current trends among pediatric ophthalmologists to decrease myopia progression: An international perspective. *Graefes Arch Clin Exp Ophthalmol 256*(12), 2457–66.
14. Nouwen, H. (2017). *You are beloved: Daily meditations for spiritual living*. Convergent.

2

FEEL

Ninety percent of infants cry at birth. Baby Ashlyn did not. Instead, she sat silently in her infant warmer, swaddled in blankets.

The cry of a newborn signals to a pediatrician that the baby is vigorous, healthy, and breathing on its own. The absence of a cry gives doctors and nurses reason for concern. Are the lungs functioning correctly? Has the baby successfully made the transition from the womb to the world? What needs to be done to ensure their safety?

Ashlyn was breathing just fine. Her heart rate was normal. There was nothing to worry about, according to the on-call pediatrician. Yet the silence was an ominous sign of things to come.

Diaper rashes are common in infants. Over time, these red, irritated areas with satellite lesions cause significant pain. But Ashlyn hardly seemed to notice the expanding fungal infection. Her parents cringed every time they cleaned the wounds around her bottom, secretly wondering why she wasn't crying. When they brought her to the doctor, he simply gave instructions to treat the site. Diaper rashes happen to most babies. Nothing to worry about.

At six months, Ashlyn's left eye became swollen and red. It looked like typical conjunctivitis to her physician, who prescribed the appropriate antibiotic drops. But when the pink eye did not improve with treatment, the family was referred to an ophthalmologist who discovered an undiagnosed corneal abrasion. Even with this painful condition, Ashlyn just sat there, happy as can be.

Worried that Ashlyn had no corneal sensation in her eyes, the eye doctor referred the child to Nemours Children's Clinic for further evaluation. While waiting for the specialist appointment, Ashlyn almost severed her tongue with her newly formed baby teeth. The ensuing months brought a comprehensive evaluation with a battery of assessments. Most of the bloodwork and radiographic exams were normal. One ominous test came back positive.[1]

It took almost two years, but specialists finally came up with the correct diagnosis: congenital insensitivity to pain. After receiving the alarming news, her parents searched the internet, which yielded few results and even less guidance. They found horror stories of mutilation and early death, like a boy in India who died after jumping off his roof. Having no sense of pain, the boy thought he was invincible.

Congenital insensitivity to pain means you cannot perceive physical pain. Inherited through a genetic mutation in the SCN9A gene, this rare disease often leads to unintended injuries and long-term health issues that can significantly shorten a person's life.[2]

In her first few years, Ashlyn accumulated countless scars, broken bones, and seared skin. When she was two, she touched the hot muffler on her father's running car when he took his eyes off her for a split second. She suffered third-degree burns but gave not a whimper. As a toddler, she ran around for several days with broken ankles before anyone noticed.

To Ashlyn's friends, she was a special classmate. They were curious about her seemingly superhuman powers. Would it hurt if someone stabbed her in the arm? How about a punch to the face or stomach? "I can feel pressure, but I can't feel pain. Pain! I cannot feel it! I always have to explain that," Ashlyn said as a teenager in a *New York Times* article about her condition.[3]

In time, Ashlyn's parents learned to watch for potential injuries before they happened. Her schoolteachers watched her like a hawk during school hours. After recess, nurses performed a daily "pit stop" to check her entire body for injuries and to wash her eyes. This ensured that no fracture or wound went unnoticed. Her longtime rheumatologist, Roland Staud, once remarked, "Her life story offers an amazing snapshot of how complicated a life can get without the guidance of pain. Pain is a gift, and she doesn't have it."[4]

EMBRACING PAIN

As Ashlyn's story shows, physical pain can serve important purposes in our lives. Though we don't often think of pain as a gift, a lack of sensation can be devastating. It may be difficult to understand at times, but God sometimes uses our pain and redeems our suffering for his purposes.

How do we sense pain? The feeling of pain results from the communication of electrical signals to our brains from nerve endings within our skin. The SCN9A gene codes for sodium channels that transport positively charged ions into cells. Once the ions are generated, an action potential transmits the signal to the dorsal horn of the spinal cord and eventually to the brain. Only then do we become conscious of pain.[5]

Nerve cells, or nociceptors, perceive various sensations within the body, including pain and pleasure. But they fire more rapidly when experiencing agony and discomfort, beckoning us to withdraw from potentially harmful stimuli. This natural response protects us from injury and harm.

Only a perfect balance of pain sensitivity—accurately detecting and sensing pain without experiencing an extreme occurrence of it—allows us to live abundantly and worship fully.[6] Mutations in the SCN9A gene result in unique disorders, from congenital insensitivity to pain, as with Ashlyn, to extreme pain disorder, which results in sudden, intense attacks of pain in various parts of the body. Patients with this disorder experience excruciating pain triggered by changes in temperature, emotional distress, and even the eating or drinking of certain foods or beverages.[7]

Whether physical, spiritual, or emotional, pain gets our attention. In a world filled with noise, pain speaks to us when we cannot hear. The more painful the experience, the longer we pause in response. The deeper the wound, the more time we need to heal.

C. S. Lewis once wrote, “God whispers to us in our pleasures, speaks in our conscience, but shouts in our pain: it is His megaphone to rouse a deaf world.”[8]

The apostle Paul speaks to this counterintuitive design, sharing in Romans 5:3–5 how “suffering produces endurance, and endurance produces character, and character produces hope.” We rejoice in our suffering because God’s love has been poured into our lives through the Holy Spirit.

The experience of pain is a hallmark of Christianity.

Whether we're mature in our faith or recent converts, suffering teaches us to depend on God. When all is well, we have no need for a savior. But in the brokenness of relationships and the death of loved ones, we realize the frailty of life. God uses pain to gain our attention and sanctify our souls.

To be sanctified means to be set apart. Sanctification is separation from sin and separation to God.[9] In 1 Thessalonians 5:23, Paul says we should be sanctified completely and be kept blameless at the coming of Christ. It is a progressive process by which we are made holy, set apart for God's purposes.

How can we be purified and made holy? Only through the pain of sanctification. Hebrews 4:12–13 describes this process vividly. There we see the Word of God penetrating joints and marrow, soul and spirit. Only with this level of scrutiny does it judge the thoughts of our minds and attitudes of our hearts. It is similar to an oncology surgeon removing clear margins of cancer. Sin is slowly and painfully cut away from our lives. What is left is the Spirit of the living God, allowing us to live pure and blameless lives.

PURPOSEFUL TOUCH

Pain is a fundamental component of sensation, but only part of the exquisite design of human touch. Our touch is the earliest sense to develop and the most mature of all the senses in utero. Arising early in pregnancy, skin sensitivity is crucial to babies' growth. Research shows that infants born without adequate sensory stimulation experience developmental delays.

In the hospital, infants often suffer from mechanosensory

deprivation. That's because premature infants are isolated in incubators for long periods while their bodies learn to regulate temperature and their vital organs grow and mature. Unfortunately, this unnatural separation can prevent the natural bonding with mothers of healthy babies.

A mere fifteen minutes of physical stimulation three times a day results in babies growing more and developing faster. Those who receive this kind of touch average 47 percent greater weight gain than unstimulated babies. Stimulated infants also spend more time active and awake while demonstrating greater maturity in their orientation, habituation, and motor skills. And even after eight to twelve months, these same infants still weigh more and score better on developmental and motor assessments.[10]

On October 17, 1995, two premature twins in Worcester, Massachusetts, were born twelve weeks before their due date. They each weighed just two pounds. At the time, the standard procedure called for the infants to be placed in separate incubators to reduce the risk of infection while fostering the maturation of their organs. Kylie, one of the twins, thrived with adequate weight gain. But Brielle, her sister, struggled to breathe and grow.

In the ensuing month, Brielle's condition worsened dramatically. Her heart rate climbed as she struggled to breathe while her limbs turned blue. Desperate to intervene, the twins' nurse, Gayle Kasparian, tried a unique approach. After obtaining parental permission, Gayle moved the twins into the same incubator. Though this was not standard practice in North America, it was regularly practiced in Europe.

In a few seconds, Brielle moved close to Kylie. One of the most iconic photographs of the year, the *Rescuing Hug*, shows Kylie wrapping her left arm around Brielle's shoulder.

This physical touch resulted in an improvement in Brielle's oxygenation. Over time, her heart rate decreased and her temperature normalized. She became less agitated and began gaining weight.[11] The gentle touch of a twin gave life to her sister.

God designed all humanity to receive physical touch. At birth, almost 20 percent of the body is skin. It is the largest of all organs, roughly 2,500 square centimeters. The adult skin grows to nineteen square feet, or 19,000 square centimeters. Millions of different cells can be found within, including five million sensory cells and 350 different types per centimeter.[12]

The whole body is covered with this flexible, contiguous structure. The skin protects us from the outside environment but also lines the insides of our mouths, noses, stomachs, and intestines. At every moment, infectious diseases, electromagnetic radiation, extremes of temperature, and changes in elemental states threaten the body. Our skin protects us from these things.

Our epidermis, the outermost layer of the skin, reflects the world around us while the inner surface communicates these external realities to the cells and organs within. In his landmark book *Touching*, anthropologist Ashley Montagu says, "The skin is the mirror of the organism's functioning; its color, texture, dryness, and every one of its other aspects, reflect our mental state and our deeper sensations. We blanch with fear and turn red with embarrassment. Our skin tingles with excitement and feels numb with shock; it is a mirror of our state."[13]

From this magnificent fabric, touch arises. It is the mother of all other senses, a precursor to the eyes, ears, nose, and mouth. The skin alone differentiates into other organs. A six-week-old embryo will bend it's neck or trunk away from light stimulation of its lips. When a baby's palm is touched, the thumb and fingers clench like a fist as early as the second trimester.

A newborn baby turns its head and opens its lips if touched in the corner of the mouth. This rooting reflex helps infants find their mother's breast or prepare to suck on a bottle. If you lightly stroke a baby's foot, the big toe moves upward while the others fan out. Even when newborns are too young to walk, they move their legs as if trying to take steps when their feet touch the ground.

The epidermis is responsible for touch. It is replete with free nerve endings and nerve plexuses. The larger nerve plexuses, Pacinian corpuscles, respond to stimulation from pressure and tension. Not surprisingly, they are numerous within the fat pads of the fingers, toes, and other delicate body areas.[14]

More than half a million sensory receptors exist in the body—a wonder of God's creative genius. Skin sensitivity comes from this tactile system of receptors that respond to temperature, pressure, pain, and pleasure. Sensory fibers communicate painful wounds or delicate touches to the brain through the spinal cord. In the cerebral cortex, we find the precentral gyrus, or convolution, which is responsible for sensory functions.

The sensory homunculus, or map of the cerebral cortex, has long represented where each sensation in the body is processed within the brain. Though perception occurs throughout the body, each sensation is *deciphered* in specific areas. The larger the area of representation in the brain, the more critical the tactile functions of these body parts for a person and their future development.

Some parts of our body have greater touch sensitivity than others. For example, the tips of our fingers can identify complex patterns and textures while our knees can't detect such fine detail. The hands and the lips play an important role in feeling, so they require more of our brains to process the information they receive through touch.[15] And because God designed our fingers, for instance, to identify such fine detail in the things they touch, people have found creative ways to use this capability to improve others' lives.

DESIGNED FOR REDEMPTION

For centuries, formal education was unavailable to the blind. Those who were unable to see were mostly helpless and were considered a burden to society. It wasn't until the eighteenth century that schools for the blind were started. The life of Louis Braille testifies to this transformation in blind education through touch. Born in 1809 in Coupvray, France, Louis was the fourth son of Simon-René and Monique Braille. As a village saddler, Simon-René made harnesses, saddles, and other horse tackle. While playing in his father's shop one day at the age of three, Louis accidentally plunged an awl into his right eye. Though only one eye was affected, sympathetic blindness quickly ensued, causing blindness in his other eye.[16]

With few options for education, Louis still managed to excel in school. Adept at listening, he learned through hearing and became a prodigious organist. By age ten, Louis was offered a scholarship to attend the National Institute for Blind Children in 1819. As the first school of its kind, the institute taught academic and vocational skills to blind children.

There, Louis met Charles Barbier. While serving in the French army, Barbier invented a code of twelve raised dots representing different sounds. This system of sonography allowed for the decoding of messages by touch. It was intended for silent communication by soldiers in the dead of night. Barbier soon realized its utility for the blind as well.

Braille was fascinated by Barbier's learning system but thought twelve dots were far too complex for coding. Even the healthiest of soldiers had great difficulty learning and implementing it. So Braille decided to refine and simplify the system for widespread use, consolidating the twelve dots to six. And instead of using a sound-based system, he assigned combinations of dots to letters and punctuation marks, yielding sixty-four total symbols.

The newly designed code was met by stiff resistance, the most significant of which was a negative societal attitude toward braille.[17] Instead of a separate system, many people believed the blind should be taught how to read printed letters. But over time, braille became the world's preeminent system of touch reading and writing. By moving their fingertips from left to right across the dots, blind individuals were now able to read. A slate and stylus allowed them to write. Today, modern computers and electronic devices streamline the reading and writing of braille.

When other senses fail, our touch can compensate in remarkable ways. For the blind, braille is life giving. Every cursory movement of the fingers and thumb perceives and processes nonvisual information to an extraordinary degree. Their heightened sense of touch is a window to the world. Though they do not possess a physical gaze, they see with their hands.

Throughout the Bible, we can appreciate the redemptive design of touch. In almost every instance, healing is accompanied by the laying on of hands. Why? There is transforming power in touch. Jesus knows this and teaches his disciples to lay their hands on others before his ascension. For those fortunate enough to be touched by the Son of God, it is healing, cathartic, and liberating.

Whom Jesus touches is just as important as what they receive. Jesus' grasp transcends the boundaries of society. He lays hands on the orphans and the widows, the deaf and the blind, the oppressed and the outcast. For the outcast, he provides hope. He gives dignity to the untouchable. In the face of certain judgment, he offers forgiveness.

In Mark 1:40–45, a leper comes near to Jesus, imploring him to heal his leprosy. His words, "If you will, you can make me clean," underline his status as an outcast in society. Lepers were considered unclean not only because of the infectious nature of the disease but also because of the appearance of their skin, which looked dead. According to Numbers 19:11, touching a dead body made a person unclean for seven days.

The disease of leprosy, caused by *Mycobacterium leprae*, often presents as loss of feeling or sensation within hypopigmented patches of skin. In those days, no treatment existed for this severe skin disease that caused peeling and scabs. Because of this, lepers were confined to living outside regular communities. They wore tattered clothes while covering the lower part of their faces. When a passerby came near, lepers shouted, "Unclean," to announce their presence. No other man or woman would touch these pariahs, yet Jesus pitied the leper. In a moment of mercy, he did what no other would, stretching out his hand to touch the afflicted person.

The touch of Christ healed the leper of his sores while likely restoring his ability to sense touch.

Reaching out to touch someone is a decision that the receiving person allows. An invitation to be touched signifies a level of trust and intimacy. The most private parts of the body are reserved for loved ones. Not only did Jesus lay his hands on the exiles and outcasts but also he allowed these strangers to touch him.

The adulterer in Luke 7 brings an alabaster jar of ointment when she learns Jesus is reclining at a Pharisee's house. Similar to lepers, promiscuous women were considered unclean by the Jews. Nevertheless, Jesus allows the woman to wet his feet with her tears and anoint them with oil. Afterward, she wipes his feet with her hair. Simon, the Pharisee, condemned Jesus for this, not realizing that his words indicted himself: "If this man were a prophet, he would have known who and what sort of woman this is who is touching him" (Luke 7:39).

Jesus teaches Simon about the gift of mercy through the invitation of touch. For most Jews, it was a serious breach of propriety for an adulterer to wash, kiss, and anoint the feet of a righteous man or priest. Yet Jesus' unblemished, sinless character precludes any thoughts of impropriety. Instead, the woman's washing, kissing, and anointing are striking acts of worship and affection for the King of Kings. With that display, according to Jesus, the woman's sins, "which are many, are forgiven—for she loved much" (Luke 7:47).

Not only does Jesus touch the lepers and outcasts but also he lays his hands on children. While the disciples were rebuking parents who brought their children to the Savior, Jesus tells them to "let the little children come to me and do not hinder them, for to such belongs the kingdom of heaven"

(Matt. 19:14). Then he lays his hands on each of them. Though he does not heal disease or sickness in this scene, Jesus understands that his touch is an encouragement and blessing to every child. What a departure from our worship services, where children are often seen as a noisy distraction.

After Jesus ascends into heaven, the disciples continue the tradition of healing through touch. Paul experiences the miracle of healing through Ananias's laying on of hands in Acts 9. Three days after he is blinded during the Damascus Road encounter, Paul's sight is restored and he embraces his calling as an apostle of Christ and missionary of the Word. He later heals a sick man in Malta in a similar fashion.

What is the source of this healing power? While we see the power of Jesus' touch throughout the Gospels, the laying on of hands in Acts signifies the presence of the Holy Spirit. The giving and receiving of the Holy Spirit are manifest through touch. Just as the mighty rushing wind and the tongues of fire are signs of the Spirit in Acts 2, the laying on of hands is the means by which the Spirit empowers Christians.

TEMPTATIONS OF TOUCH

While the laying on of hands can confer the extraordinary power of the Holy Spirit to believers, we must be equally careful with what we touch. Our hands can serve as a tremendous source of encouragement, blessing, and healing. Yet they can just as easily become instruments of destruction and sin.

Solomon describes in vivid detail the beauty of a woman's body. In Song of Solomon 7, he describes his lover's thighs as jewels, the work of a master's hands. Her navel is a rounded bowl, never lacking wine. Her neck is an ivory tower; at the top lie eyes reflecting images like pools in Heshbon. Flowing locks of hair encircle her head while her nose stands like a tower. In the covenant of marriage, the body is meant to be seen by the eyes and touched by the hands.

But when this magnificent creation is coveted by sinful eyes and possessed by evil hands, it leads to death. The Hebrew word *naga* means "to touch." In Genesis 20:3–7, King Abimelech covets Sarah and moves to add her as one of his wives. Unbeknownst to him, she is already married to Abraham. But God in his infinite mercy does not allow Abimelech to touch Sarah, saving the king from certain death. Similarly, Boaz protects Ruth from the *naga* of young men while allowing her to glean from his fields (Ruth 2:9). Touch is defiled when done with improper hands and an impure heart. Proverbs 6:29 severely warns those who commit adultery with their neighbors' wives. None who touch in this way will go unpunished.

In the New Testament, the Greek word *haptomai* has a meaning similar to the Hebrew word *naga*. Sexual implications are not always intended within the thirty-five appearances of the word. But the greatest caution exists with sexual touch. In 1 Corinthians 7:1 Paul uses *haptomai* in asserting that it is "good for a man not to touch a woman" (NASB). Some translations (ESV, NIV) consider it a colloquialism, saying a man shouldn't have sexual relations with a woman. Either way, it is not a prohibition of sex within the context of marriage, which Paul continues to encourage. The translation of *haptomai* in 1 Corinthians 7:1 as "touch" leaves open the ambiguity of a range of good expressions, including sexual relations.

In Colossians 2:21, Paul rails against strict regulations not to handle, taste, and touch. For the Colossian church, issues of rigorous asceticism were in question. Legalistic church members judged individual choices of food and drink, promoted self-abasement practices, and exaggerated bodily discipline. Paul's words reflect his opposition to an extreme repudiation of the flesh, yet caution against overindulgence.

That is the key to our sense of touch and the way it informs our spiritual health. Touch was created as a blessing for us and for those with whom we come in contact. The same hands are used for eating, drinking, and working. They exert tremendous influence over what and who we grip. Touch can be destructive and unholy, like Abel's killing his brother, Cain. Or it can be life giving and redemptive, like Jesus' healing touch.

Perversions of touch are particularly disturbing. Physical and sexual assault destroy God's intent for our hands and fingers. When committed against minors, these acts steal innocence and youth. Nevertheless, this deviant touch does not render an abused person unclean. Their spiritual status before God is forever unchanged. His desire is for their healing, both physically and spiritually.

Will we choose to be sanctified in the way we use touch, or allow our hands to turn to sin?

ACTS OF WORSHIP

The sense of touch contributes to our sanctification and faith. Our nerves allow us to feel God while experiencing his creation. A firm handshake or gentle touch speaks encouragement to those we love. Laying on of hands is a conduit for healing the sick and brokenhearted. Every movement of the fingers and thumb can be an act of worship of the living God.

Our touch allows us to feel the physical presence of God. We remember Jesus' sacrifice while partaking of the sacrament of communion. Jesus invited his disciples to take the bread and cup during the Last Supper. Just as they felt the texture of the bread and the shape of the cup, our hands come into contact with his body broken for us and the blood of the new covenant, just as Jesus' disciple Thomas did. Some believe the bread and wine are physically changed into the body and blood of Christ. To others, they are merely symbols of Christ's death. In either circumstance, our sense of touch helps facilitate this holy communion.

When a believer is baptized, they participate in the life, death, burial, and resurrection of Jesus. It is a sign of a Christian's initiation into the life of the church and a public declaration of their faith in and submission to Jesus as lord and savior. Though different traditions debate the theological implications of this sacrament, water is a constant for all baptisms.

Just as the earliest Christians could feel the cool water against their skin while being baptized in ancient lakes, rivers, and pools, we also sense the cascade of these life-giving molecules of liquid on our nerve endings. When immersed, we find the perfect image of death. For a brief moment, we cannot breathe while holding our breath. But in an instant we arise from the water and feel it dropping away, signifying the resurrection of Christ. Water is the perfect physical expression of baptism for our sense of touch.

Among the fellowship of believers, gentle physical

contact demonstrates brotherly love in ways words do not convey. In 2 Corinthians 13:12, Paul encourages us to greet one another with a holy kiss. For a variety of reasons, phobias, and concerns (or self-concern), too seldom do Americans engage in these potentially edifying behaviors. In the absence of physical contact, we are deprived of something God created to bless our communities of faith.

Observations by psychologist Sidney Jourard provide evidence of differing levels of physical contact among cultures. In the United States, two friends sitting together touched each other twice an hour. In France, the number increased to 110 times every hour. And in Puerto Rico, friends touched one another an astounding 180 times per hour.[18]

As Christians, we would do well to encourage brothers and sisters with a warm handshake, a gentle pat on the back, or an endearing hug. A welcoming touch comforts and soothes while signaling trust and love. When initiated with a pure heart, we can bless others in a manner above reproach—even with a holy kiss.

When there are sick and suffering among us, we should heed the words of James 5:16. The anointing of oil coupled with the laying on of hands allows the power of the Holy Spirit to flow through our fingers. Just as healing occurred through the disciples and early Christians in Acts, so too can the prayer of faith usher in God's healing for the sick in today's age. There is great power in the prayer of righteous people, as James proclaims.

Whether or not physical healing occurs for the ill in this present life, spiritual healing is available at all times. Those who are touched by the hands of brothers and sisters can also be moved by the presence of the Holy Spirit while experiencing the love of the Father. That's why anointing a pastor for ordination or consecrating an elder or a deacon usually involves laying on of hands. It is more than symbolic. The physical touch of hands coupled with prayer is transformative.

Every stroke of the hand can be an act of worship or blessing to others. Our touch reflects our intimacy with God and empathy toward our neighbors. May we be ever mindful of the remarkable power touch possesses, and may our hands reflect Jesus' selfless touch in all things.

QUESTIONS FOR REFLECTION

1. Congenital insensitivity to pain reminds us how precious the sensation of touch is. How do you take for granted this sense of feeling?
2. How has modern society perverted the sensation of touch and feeling?
3. What was God's design for the organs of touch?
4. A holy kiss or firm handshake can be redemptive. How can you bless others with the gift of touch?
5. In what ways can you experience Jesus through your sense of feeling?

NOTES

1. Heckert, J. (2012, November 15). The hazards of growing up painlessly. *New York Times Magazine*. https://www.nytimes.com/2012/11/18/magazine/ashlyn-blocker-feels-no-pain.html; Trachtenberg, T. (2012, July 5). The girl who can't feel pain. ABC News. https://abcnews.go.com/blogs/health/2012/07/05/the-girl-who-cant-feel-pain

2. Genetic and Rare Diseases Information Center. (n.d.). Congenital insensitivity to pain-anosmia-neuropathic arthropathy. National Institutes of Health. Retrieved May 30, 2022, from https://rarediseases.info.nih.gov/diseases/12267/congenital-insensitivity-to-pain
3. Heckert, J. (2012, November 15). The hazards of growing up painlessly. *New York Times Magazine.* https://www.nytimes.com/2012/11/18/magazine/ashlyn-blocker-feels-no-pain.html
4. Heckert, J. (2012, November 15). The hazards of growing up painlessly. *New York Times Magazine.* https://www.nytimes.com/2012/11/18/magazine/ashlyn-blocker-feels-no-pain.html
5. SCN9A sodium voltage-gated channel alpha subunit 9. (n.d.). National Center for Biotechnology Information. Retrieved 2022, from https://www.ncbi.nlm.nih.gov/gene/6335
6. Genetic and Rare Diseases Information Center. (n.d.). Congenital insensitivity to pain-anosmia-neuropathic arthropathy. National Institutes of Health. Retrieved May 30, 2022, from https://rarediseases.info.nih.gov/diseases/12267/congenital-insensitivity-to-pain
7. Genetic and Rare Diseases Information Center. (2021). Paroxysmal extreme pain disorder. National Institutes of Health. Retrieved July 9, 2022, from https://rarediseases.info.nih.gov/diseases/12854/paroxysmal-extreme-pain-disorder
8. Lewis, C. S. (1940). *The problem of pain.* Harper Collins.
9. Simpson, A. B. (1890). *Wholly sanctified.* Christian Alliance Publishing.
10. Field, T. M. et al. (1986). Tactile/kinesthetic stimulation effects on preterm neonates. *Pediatrics* 77(5), 654–58.

11. The hug that helped change medicine. (2013, February 22). In *The Situation Room with Wolf Blitzer*. https://www.youtube.com/watch?v=0YwT_Gx49os
12. Bolognia, J. L., Schaffer, J. V., & Cerroni, L. (2018). *Dermatology* (2 vols.). Elsevier.
13. Montagu, A. (1971). *Touching: The human significance of the skin*. Columbia University Press.
14. Bolognia, J. L., Schaffer, J. V., & Cerroni, L. (2018). *Dermatology* (2 vols.). Elsevier.
15. Louis, E. D. et al. (2022). *Merritt's neurology*. Wolters Kluwer.
16. Kugelmass, J. A. (1951). *Louis Braille: Windows for the blind*. Messner.
17. Riccobono, M. A. (2006). *The significance of braille on the blind: A review and analysis of research based literature* [unpublished paper]. Johns Hopkins University.
18. Jourard, S. M. (1966). An exploratory study of body-accessibility. *Br J Soc Clin Psychol 5*(3), 221–31.

3

SMELL

Whenever I visit botanical gardens, I am drawn to the fragrant plants. The sight of beautiful flowers filling nature with depth and color is always breathtaking. Yet the added scents of lavender, rose, honeysuckle, orange blossom, and magnolia tickle the sense of smell. Hundreds of chemical compounds give flowers their distinct fragrances. Each is so enticing that perfumers spend decades studying botany and designing scents based on the oils from these flowers.

We appreciate the sense of smell overtly at times—when a fragrance is particularly sweet or an odor particularly offensive. But most often, we experience smell subconsciously. Its wonders enhance our world—until it's gone. Some are born with anosmia, a partial or total loss of smell. Others experience it after suffering illness with a virus such as the SARS-CoV-2 virus. Although not life-threatening, anosmia can profoundly affect daily life. Most people with loss of smell experience spontaneous recovery from their symptoms within six months. For others, anosmia lingers much longer.

When it does, rose oil's warm, floral redolence disappears. We miss the poetry of freshly turned soil or newly mown lawns, the heady experience of a sweetheart's cologne, the incomparable smell as we hold a newborn close. After losing her sense of smell, one young mother said this of the candles her mother sent for her birthday: "Once rosemary and lemon balm, were now nothing and nothing."[1]

Our sense of smell is also a guardrail. Spoiled milk. Soiled diaper. Smoke. Hot asphalt. The musky scent that shows up before a bear crosses our path.

Smell is purposeful, protective, and life enhancing.

Olfaction is closely tied to our appetite. Many who lose their sense of smell find their ability to taste hampered. Unsurprisingly, those who lose the ability to smell may succumb to anhedonia, the inability to feel pleasure. Social isolation and detachment ensue when food and wine are no longer enjoyable. In a study of people affected with sudden-onset anosmia, one patient shared, "I feel alien from myself. . . . It's also kind of a loneliness in the world. Like a part of me is missing, as I can no longer smell and experience the emotions of everyday basic living."[2] No wonder olfactory dysfunction is often associated with depression and anxiety.

For those who regain their sense of smell, parosmia may linger. That's when the sense of smell is distorted. It most often manifests with delightful scents morphing into pungent malodors. The once alluring fragrance of a perfume becomes intrusive and nauseating. Instead of a lightly caramelized, almost nutty roast, coffee smells rancid. Some describe their favorite fragrances—toast, coffee, bacon—as smelling like industrial waste.

Our memories tell us what something should smell like. They draw us back to our initial encounter with food, wine, or drink. Each of these past experiences causes us to smile with joy or cringe in disgust. But with parosmia, when we sniff the air, our brains register a completely different scent. Memories of old are incongruent with present experience.

THE NOSE KNOWS BEST

As a child, I always looked forward to my mother baking in the kitchen. As she added ingredients to a bowl and blended the batter, I noted the sweet aromas of butter, sugar, vanilla, and spices in the air. As pastries and cakes slowly baked in the oven, they filled the kitchen with a delightful fragrance.

The mouth-watering aromas of chocolate brownies and cream-cheese pound cakes still bring me back home. My senses awaken each time my nose inhales these heavenly scents. To this day, the aromas remind me of my childhood—of a young boy eagerly awaiting freshly baked items to pop out of the oven.

How do our brains process the delightful scents of French pastries and Italian desserts? People often think of smell as the weakest of the senses. But nothing could be farther from the truth. The differentiation of innumerable odors shows the remarkable plasticity of smell. Yet there is a simplicity in the way God created the complex olfactory system.

Smells begin their journey in our bodies high inside our nostrils. Particles enter either directly through the nares or across a channel connecting the roof of the mouth with the nose. Millions of olfactory sensory neurons reside in this specialized patch of tissue. Each sensory neuron has but one receptor. Three hundred to five hundred different odor receptors exist, much fewer than the countless odors available in nature.

How do we differentiate between smells with only a few hundred smell receptors? Simple odors trigger particular groupings of olfactory neurons—their unique pattern codes for messages communicated directly to the brain. Since the message is instantaneous, we know immediately whether something is fragrant or fruity, sweet or savory.

What about more complex scents? Your favorite roasted coffee alone contains about eight hundred to nine hundred volatile odor molecules. That's three times the number of smell receptors in our bodies. With more complex scents composed of different compounds, one odor can activate more than one neuron and alter that cell's response to another smell. These interactions increase or decrease the strength of that response, allowing our bodies to process countless different scents.[3] It's comparable to the highly complex and changing images on television screens. The patterns of three colors (red, green, blue) combine to represent an almost limitless number of other images and colors.

For centuries, conventional wisdom assumed the nares could differentiate only approximately ten thousand smells. But Rockefeller University researchers proved the nose might be even more gifted than the eyes in processing stimuli.[4] The research tested 128 molecules in beverages to see how well people could distinguish the odors. Molecules were randomly mixed in groups of ten, twenty, and thirty to make unfamiliar aromas. Three samples were given for people to identify, two of which were the same and one which was different.

Most participants could correctly distinguish smells if the ingredients varied by more than 50 percent. No one could discriminate odors that overlapped by more than 90 percent. Based on the collective answers, the researchers modeled how many different scents a typical person could identify. For an average person, it was more than one trillion unique smells! Even the olfactory-challenged could still smell eighty million variants.

FROM PERFUMES TO ESSENTIAL OILS

Perfumes have long captured the imagination. From ancient times to modern days, they are sprinkled on for daily use and bought for special occasions. The world's most iconic perfume—Chanel No. 5—is described as a "floral bouquet, composed around May rose and jasmine, [featuring] bright citrus top notes."[5] In creating his masterpiece, perfumer Ernest Beaux used for the first time an artificial synthetic aldehyde, blending various florals to give a unique composition. Eighty different ingredients in the perfume include "rose, ylang-ylang, jasmine, lily of the valley, and iris—all layered over a warm, woody base of vetiver, sandalwood, vanilla, amber, and patchouli."[6]

Globally, more than seventeen thousand perfumes are available to consumers. Each contains between fifty and three hundred ingredients. In ancient times, only a couple of hundred raw materials were readily available. Each was derived from natural sources, primarily plants. Today, synthetic compounds formulated in the organic chemistry lab offer us countless new scents.

A professional perfumer must accurately identify every compound in their formulary, sometimes up to several thousand items. How do they do this? The Givaudan method, invented by Jean Carles, is the leading technique for training novice perfumers. Students learn to differentiate families of fragrances, such as woody, floral, amber, and fresh, with a matrix approach. Over time, they begin to recognize the subtle differences within families. For example, citrus scents become more specifically lemon, mandarin orange, lime, grapefruit, and tangerine. Perfumers then learn to recognize accords, which are combinations of ingredients that mix well. The most recognizable perfumes are composed by mixing several different accords.

Even in Old Testament antiquity, treasured perfumes were used as anointing oils. Exodus describes the formula for these holy emollients when the Lord says to Moses, "Take the finest spices: of liquid myrrh 500 shekels, and of sweet-smelling cinnamon half as much, that is, 250, and 250 of aromatic cane, and 500 of cassia, according to the shekel of the sanctuary, and a hin of olive oil. And you shall make of these a sacred anointing oil blended as by the perfumer; it shall be a holy anointing oil" (Ex. 30:22–25).

The earliest apothecaries blended ingredients such as myrrh, cinnamon, aromatic cane, cassia, and olive oil precisely to create delightful aromas. The cost of these ingredients was substantial. Each blend anointed sacred places of worship like the temple, altar of incense, and ark of the testimony.

Every person and object touched by these essential oils was considered holy, set apart. Anointing the priest, temple, and other artifacts meant they received divine protection from God. Yahweh himself directed the Israelites to manufacture these precious oils to consecrate his sanctuary and ordain his chosen servants.

Since the dawn of creation, essential oils have existed

on earth. They are natural emollients containing volatile chemical compounds. Typically derived by distillation, these oils emit characteristic fragrances of the plants they are extracted from. Each scent is unique, just like the properties of the emollient.

In modern times, essential oils have been linked to better sleep, alleviating headaches, and even helping with sore throats. But do these concentrated plant-based oils have health benefits? Although many people may consider essential oils as a treatment for numerous ailments, there is a paucity of evidence to determine their clinical efficacy. Some lab studies are promising, yet others show no symptomatic improvement.[7]

The Bible references more than thirty-three different essential oils. Some are highlighted individually; others are combined to form perfumes, incense, and anointing oils. They were used for medicinal purposes, cleansing people and homes, honoring and showing devotion, and soothing ointments. Four of the five ingredients of the holy anointing oil of Exodus 30 are considered essential oils: myrrh, cinnamon, cane, and cassia.

Myrrh is cited more than 150 times in Scripture. Its cooling scent has been described as sweet, aromatic, and woody. Most commonly, we see it used in anointing God's servants. But myrrh is also an ingredient of incense, used for embalming, and was considered by Queen Esther to be a beauty ointment. As the magi knelt before a newborn Jesus, they offered gifts of gold, frankincense, and myrrh. After his crucifixion, Jesus was wrapped in a mixture of myrrh and aloe.

Perhaps the most recognizable essential oil is cinnamon. The sweet yet spicy scent is commonly used for cooking and baking. A potent antifungal, it is also a necessary ingredient in air fresheners. In Proverbs 7:17, a perfume blended with cinnamon, myrrh, and aloe was King Solomon's choice for his bedroom.

Though not an essential oil found in Exodus 30, frankincense is one of the most famous scents in the Bible. It was a vital ingredient for making incense but also held high medicinal value. Christians likely recognize it from the birth of Jesus. On that auspicious day, it was worth just as much as gold.

If we ever wonder about the importance of fragrances, the numerous references to their presence and properties recorded in the Bible should refute any doubt. Smell is not only a divinely created sense but also a gift woven throughout Scripture and daily life.

A HEIGHTENED SENSE OF WORSHIP

Helen Keller once wrote, "Smell is a potent wizard that transports you across thousands of miles and all the years you have lived. The odors of fruits waft me to my southern home, to my childhood frolics in the peach orchard. Other odors, instantaneous and fleeting, cause my heart to dilate joyously or contract with remembered grief. Even as I think of smells, my nose is full of scents that start awake sweet memories of summers gone and ripening fields far away."[8] Whether awakened by a fragrant lily or a pungent odor, our sense of smell carries us back to specific moments in time.

In the brain, the olfactory system is connected to the hippocampus. Embedded deep within the temporal lobe, the hippocampus is a complex structure that plays a significant role in learning and memory. It is vital for encoding memories and consolidating them. If it's damaged by injury or illness, a person's ability to form new memories may be compromised, like the protagonist in the movie *Memento*.

Memories allow us to worship God in the present as we look to the past. Whether of the time you accepted Jesus as savior or of the most recent sermon that touched your heart, memories of the Spirit's work in our lives connect us with the Father. These past deeds foster gratitude while inviting dependence and servitude. During the Last Supper, Jesus invited the disciples to partake of the bread and cup in remembrance of him. The very discipline of remembering is an act of worship.

Through intracranial electrophysiology and neuroimaging, scientists have shown that the sensory system most highly linked to the hippocampus is the olfactory one.[9] Compared with sight, taste, sound, and touch, smell triggers the recollection of latent memories most powerfully.

Smells that pleased us as children can continue to calm us as adults in times of stress and turmoil.[10] Whenever I am anxious, the sweet scent of freshly baked chocolate-chip cookies still soothes my soul while bringing me back to my childhood. Likewise, odors that once caused fear and trepidation can incite unpleasant feelings for many years. Every time I smell the foul odor of locker-room sweat, I recoil at the memory of being cut from the high school varsity basketball team.

In the Bible, every scent of perfume, essential oil, and incense is tied to an act of worship. Why? By etching memories into our souls, they enhance our present and future communion with God.

Though science is just beginning to unravel the nuances

behind the olfaction-memory connection, God designed our sense of smell to heighten our acts of worship. On the Day of Atonement, coals of fire were taken from the altar and combined with sweet incense to smoke over the mercy seat of the ark of the testimony (Lev. 16:12–13). Over time, Israelites associated the fragrant smell of incense with the call to worship with reverence and awe.

According to Song of Solomon 1:3, whenever the fragrance of anointing oil was in the air, the name of the Lord was ointment poured forth. The essential oils mentioned earlier—frankincense and myrrh—were symbolic, prophetic gifts (from the magi to the Christ child) and played a role in daily worship as well as significant events, anointing the altar and God's people. Their fragrance reminded the nation of Israel—and us today—of the animal sacrifices of the Old Testament. But more important, they beckon us to remember the atonement of Christ that offers hope to all generations.

Olfactory scents are not merely delightful fragrances. They are connected with purification, cleansing, and healing. After committing adultery with Bathsheba, King David pleaded to be cleansed with hyssop and washed whiter than snow (Ps. 51:7). The sweet mint scent of that simple garden herb represented purification for David, even for the most egregious of sins.

What smells bring you back to specific experiences of the past? Some you will recall with laughter and joy, and others may bring sadness and tears. Consider not only the incident and people involved but also how God was present in these moments. In what ways did he speak to you? How did he redeem?

A PLEASING AROMA TO THE CREATOR

Fragrances help deliver us into the heart of worship, and they please the Lord. When Noah offered burnt offerings after the flood subsided, God smelled the aroma of the sacrifice, and it pleased him. From then on, he promised never to curse the ground because of man (Gen. 8:21).

Though some consider the idea of the divine "smelling" an anthropomorphism, our sense of smell is predicated on the Creator's design. We know God hears—most people interpret that literally—but that's not predicated on his having physical ears. He walks and talks, but until God was made flesh in Jesus, was that with feet and a mouth? We may never fully understand the attributes of God, but it is clear Yahweh appreciated Noah's offering. The Hebrew phrase *nihoah reah* is translated as "pleasing aroma." Though it is referenced more than forty times in the Old Testament, Genesis 8:21 is the only instance where God is recorded to smell. Still, throughout the Old Testament, various sacrifices were offered to God for the atonement of sins. Pleasing aromas included burnt offerings, food offerings, and drink offerings.

Burnt offerings consisted of one young bull, one ram, and seven male lambs without defect (Num. 29:2). As these animals burned in the fire, unique scents were scattered. Though this pleasing smell wasn't the primary reason God was honored, the searing of animal meat was crucially tied to repentance. Even apart from burnt offerings, the rising smoke offered a constant reminder of the need for atonement. Incense was to be burned continually before the Lord, making a sweet savor.

If a perfumer can differentiate a trillion odors, then an omnipotent God must be able to differentiate an infinite number of scents. But what made the fragrances recorded in Scripture pleasing to the Lord? It wasn't the smell of burning incense or the scent of fine wine alone. The prophets warned that fragrant sacrifices alone would not please the Lord.

Psalm 40:6–8 tells us that God does not always delight in our sacrifices or offerings. They are made fragrant and pleasing when we delight to do the will of God, when his law is on our hearts. False worship and inauthentic sacrifice displease the Lord. In these cases, God instructs us to bring no more vain offerings. Their associated incense is an abomination to him (Isa. 1:13). Instead, he desires broken and contrite hearts in repentance.

The most pleasing aroma to God is the death of his Son. Jesus' ultimate sacrifice is considered by the Father a "fragrant offering and sacrifice to God" (Eph. 5:2). The fragrance described in that verse wasn't for our sakes but for the Father's. By giving himself up for the world, Jesus was the most fragrant offering to God. We can honor him with tithes of gold, frankincense, and myrrh. But more important, we honor him when we spread the fragrance of his life, death, and resurrection.

INSIDIOUS ODORS ARE LURKING

Only half of us can smell the bitter-almond odor of hydrogen cyanide. This colorless, highly poisonous gas that interferes with respiration prevents cells from using oxygen. Death can occur within two minutes of a person inhaling just seventy milligrams of this lethal gas.

Hydrogen cyanide was first used as a pesticide in the late 1800s. Chemists at Degesch chemical corporation added an absorbent stabilizer and eye irritant to the poisonous gas. In the presence of air, the resulting compound reacts to release hydrogen cyanide. Sealed in metal containers, the new product was marketed as Zyklon B. These innocent-looking canisters became the preferred mode of extermination of more than one million Jews during the Holocaust.

Each day during the height of World War II, trains brought countless Jewish people to concentration camps like Auschwitz. The Jews and Jewish sympathizers were sent directly to the gas chambers if deemed unfit for work.

This horrific doctrine of humane killing was predicated on victims being unaware of their fate. Every day, guards led unsuspecting Jewish elderly, disabled, and young to large rooms with bathroom tiles and showerheads, the nozzles of which were not connected to water pipes. The people were encouraged to disrobe and prepare to wash. But after the doors were sealed, an orderly opened a vent from the chamber's roof and poured Zyklon pellets into the room.

The slight odor of almonds in the air was a harbinger of death. Within moments the pellets turned into poisonous gas. Inside, frantic prisoners gasped to breathe as they pushed and shoved each other in a futile attempt to escape. With no way out, their fate was sealed. After a few minutes, everyone inside the sealed chamber was dead. After some time, the deadly gases were pumped out of the room, and the scent of fresh air replaced the stench of dead bodies.

Blood agents and toxic chemicals poisoning the body through the blood smell unique. Most are colorless volatile gases with faint odors. Arsine has a mild, garlicky smell. The distinctive stench of rotten eggs tells you hydrogen sulfide is burning. Everyone can identify the scent of bleach from chlorine found in most bathroom cleaners; it was first used in World War I.

Those distinct odors are detectible through our creator's invention of our olfactory system by design. Bad smells often serve as a warning that something is awry. The reek of rancid milk can be detected with a whiff. When food smells rotten, it likely is. Pungent odors signify the growth of bacteria and mold. Distinctive smells come from the decomposition of food or directly from invading microbes themselves.

In winter, mice always scurry into our attics to keep warm. To get rid of them, we set glue traps in opportune places. A peanut butter smear on the sticky boxes entices them into the trap. The savory aroma works every time. It's challenging to monitor every mousetrap, but over time the stink of death is unmistakable, and we're alerted that it's time to dump the glue traps and set up new rodent buffets.

Though a smell test isn't 100 percent accurate—some chemicals and microbes are odorless—it is a remarkable safeguard. God designed our olfactory system to identify dangers, whether easily recognizable or lurking in the shadows.

Illicit drugs emit characteristic odors. The stimulant methamphetamine smells like powerful cleaning products, sulfur, and burnt plastic. After addicts overdose, their sweat smells like ammonia. Those who abuse heroin secrete a sour, vinegarlike odor. With the legalization of marijuana, it's easy to recognize the musky smell of cannabis that lingers on personal items.

Because of nicotine, the smell of cigarette smoke increases a person's desire to smoke, which leaves the reek of tobacco on their breath and all over their clothes. Despite those lingering odors, the power of the nicotine makes the addictive behavior difficult to break.

The fragrance of death is not only the stench of decaying bodies but also odors of habits that lead to increased mortality. Some are obvious. A few whiffs tell us we're in danger. Others are more subtle, sometimes even inviting.

Vaping has become a trend among youth and young adults. To simulate tobacco smoking, electronic cigarettes produce vapor for inhalation. E-cigarettes don't smell the same as combustible tobacco. Manufacturers create enticing fragrances like vanilla, mint, and candy to disguise their harmful effects. A telltale sign of potential abuse is when we sense the sweet aroma of these vapors without finding the typical sources. Care must be taken to distinguish between those that foster life and those leading to death.

EMBRACING OUR SENSE OF SMELL

In 2 Corinthians 2, Paul declares that "we are the aroma of Christ to God among those who are being saved and among those who are perishing, to one a fragrance from death to death, to the other a fragrance from life to life" (2 Cor. 2:15–16). The verse doesn't say, "For we smell like the aroma of Christ," but rather says, "We are the aroma of Christ." We become. We are evidence. When a Christ follower is metaphorically sniffed, others recognize the scent of Jesus!

The stench at Auschwitz was the same odor at Lazarus's tomb. Corpses reek within a day or two. Lazarus's body had been decomposing in his tomb for four days (John 11:1–44). When Jesus arrived at Bethany, Mary lamented that he had not come soon enough. As the stone was rolled away from the tomb, Martha worried the odor would be unbearable. Yet Jesus allayed her fears, saying, "Lazarus, come out" (v. 43). As Lazarus walked forth, the stench of death was transformed into the fragrance of life.

Every day, we're presented with opportunities to redeem our olfactory senses for the work of the Lord. When we awaken, the fresh scent of a dewy morning points us to the beauty of creation. His love and mercy are new every morning. In the evening, night-scented plants release intense fragrances to attract nocturnal pollinators. As we lie down, their gentle scent reminds us of the faithfulness of God.

When we visit the beach, it's impossible not to notice the wonderful smell of the ocean. Families spend days by the shore in the summertime. Their children frolic in the waves for hours. Whether you're near the Pacific Ocean or the Atlantic, the smell of the ocean is universal. Fresh ocean air is always better than the stale smell of indoor air. It reminds us of who fills our sails.

By noticing and heightening our olfactory senses, we become more in tune with the one who created them. Just as the Lord delighted in the pleasing aromas of sacrifice, we can enjoy a multitude of fragrances he ordained for us. In ancient times, the community appreciated the sweet smell rising from open-air fire pits when the Israelites baked bread. We too can share in the wonder of a trillion scents, both natural and artificial.

Of all the senses, the olfactory system is arguably the most plastic. Most humans are born able to recognize countless smells. But we can't differentiate subtleties very well or categorize them without practice—increasing our smell agility, if you will. Experts train themselves by refining their olfactory senses with observational skills. Just six weeks of training can lead to increased thickness in several parts of the brain, resulting in increased capacity to smell.[11]

Though we may never aspire to be perfumers, imagine the possibilities if we paid more attention to our olfactory glands.

Instead of ordering black tea in the afternoon, we can enjoy the fresh and fragrant aroma of white tea-leaf blossoms from jasmine flowers.

Grilling becomes more than just eating barbecued meat. We can enjoy the rich, sweet, intoxicating fragrance of smoke before eating. Smoking meats and other food ignites the Maillard reaction. This occurs when dry surfaces are scorched with heat, breaking amino acids and sugars. Seared steaks demonstrate this char with their characteristic smell. In ancient times, smoking was used to preserve food and prevent bacteria from causing disease. Today, it is just as important in the unique scent of smoked food.

With a deeper appreciation for the smells surrounding us, our olfactory glands become an integral part of worship. Because of Jesus' sacrifice, burnt offerings and sin offerings are no longer required. In the upper room, Jesus took the bread and cup, instructing his disciples to "do this in remembrance of me." Whenever we partake of communion, the smell of wine and the aroma of bread aid us in remembering his death. When we associate these smells with his sacrifice, we pivot our thoughts from the worldly to the eternal.

Today, the smell of Advent is easy to recognize. From the release of terpene chemicals, Christmas trees emit a fresh, woodsy aroma. Conifers also release citrus, mint, and thyme into the air.

The smell of Advent candles accompanies the spicy,

fruity combination of clove-studded oranges. On a cold winter night, there's nothing like the sweet, comforting scent of hot chocolate sprinkled with cinnamon. The sweet, buttery, and spicy smell of gingerbread houses ignites the taste buds of all children. Chestnuts roasting over an open fire invite us to breathe in the magical scent of Christmas.

By associating the fragrance of Christmas with the birth of Christ, we worship the King. The smells of Advent correspond not only to his birth but to a longing for his return. Even in the times farthest from Christmas, these strong fragrances stir our hearts and minds to remember his love for us.

When we share that love with others, we become the aroma of Christ.

Many scents have a unique characteristic: they linger. Just imagine someone walking into your room after spending time at a campfire. What we smell like discloses who we've spent time with or where we've been. If we are in close communion with the Father, the Son, and the Holy Spirit, we radiate the sweet fragrance of the Word. When we are selfish and self-absorbed, we reek of the foul odors of the world.

According to Paul, we are to "walk in love, as Christ loved us and gave himself up for us, a fragrant offering and sacrifice to God" (Eph. 5:2). We sense the aroma of Christ most profoundly when we live sacrificial lives for others as Christ did for us. Serving others is a testimony to the King of Kings and Lord of Lords.

We proclaim the love of Christ through our actions and our words. In this way, it is diffused to all nations and people. There is power in the aroma of Christ. It is a fragrance from life to life for those who believe. Yet it is also a fragrance from death to death for those who will perish.

Our calling is to ensure all can smell the sweet aroma of Christ. By redeeming our sense of smell while incorporating it into worship and evangelism, we become a fragrant offering to the Lord.

QUESTIONS FOR REFLECTION

1. Everyone is drawn to particular fragrances while being repulsed by other smells. Which ones elicit these responses for you?
2. Essential oils and perfumes have existed since antiquity. According to Scripture, what role do essential oils and perfumes have?
3. Our olfactory system is extraordinary in design and function. What smells warn you that something is awry?
4. In what ways can your sense of smell facilitate worship of the living God?
5. How can you be a pleasing aroma to the Lord?

NOTES

1. North, A. (2021, December 11). My year of smells. Vox. https://www.vox.com/the-goods/22783776/long-covid-loss-of-smell-taste
2. Burges Watson, D. L. et al. (2021). Altered smell and taste: Anosmia, parosmia and the impact of long Covid-19. *PLoS One 16*(9), e0256998.
3. Xu, L. et al. (2020). Widespread receptor-driven

modulation in peripheral olfactory coding. *Science 368*(6487).

4. Bushdid, C. et al. (2014). Humans can discriminate more than 1 trillion olfactory stimuli. *Science 343*(6177), 1370–72.
5. Chanel. (n.d.). Chanel no. 5. Retrieved July 12, 2022, from https://www.chanel.com/us/fragrance/p/125530/n5-eau-de-parfum-spray
6. Helbig, C. (2022, December 1). Review of Chanel no. 5 perfume: Is it worth the hype? Byrdie. https://www.byrdie.com/chanel-no-5-review-of-chanel-no-5-perfume-346120
7. Feng, J. (2018). Identification of essential oils with strong activity against stationary phase borrelia burgdorferi. *Antibiotics* 7(4), 89.
8. Keller, H., & Shattuck, R. (2003). *The world I live in.* Publishers Group West.
9. Zhou, G. et al. (2021). Human hippocampal connectivity is stronger in olfaction than other sensory systems. *Prog Neurobiol 201*, 102027.
10. Warrenburg, S. (2005). Effects of fragrance on emotions: Moods and physiology. *Chem Senses 30 Suppl 1*, i248–49.
11. Seubert, J. et al. (2013). Orbitofrontal cortex and olfactory bulb volume predict distinct aspects of olfactory performance in healthy subjects. *Cereb Cortex 23*(10), 2448–56.

4

SAVOR

The wine was very deep, dark, and richly colored with a unique exotic nose of oriental spice, black truffles, and plenty of substance underneath—very rigorous! On the palate a full, unbelievably concentrated, sturdy wine. Chewy, intense, and with not a trace of age—this wine will last forever!" So writes Pekka Nuikki, editor of the *Fine Wine* magazines and *Fine Champagne* magazine, about the red wine Romanée-Conti 1945.[1]

The Abbey of Saint Vivant originally owned the French vineyard of Romanée-Conti in Vosne in the thirteenth century. French monks cultivated the grapevines for several hundred years before selling the land to the Croonembourge family in the 1600s. A short while later, the Prince of Conti purchased the vineyard, adding his name to it. Today, the Duvault-Blochet household operates the hallowed grounds.

The Burgundy vineyards are located on some of the world's best wine-producing soil in eastern France. Unlike the surrounding land, Domaine de la Romanée-Conti (DRG) is protected as a UNESCO World Heritage Site. Its wines are designated Grand Cru, the highest quality and most well-respected classification for wines.[2]

As the most expensive wine ever sold, a bottle of Romanée-Conti 1945 once commanded a price of more than half a million dollars. What makes the palates of sommeliers water in the presence of this delightful wine? Wine connoisseurs say it's a timeless taste of perfection straight from the heavens.

Sommeliers describe wines with the utmost precision in language. They speak of sight, color, aroma, and palate. But they also describe vinosity, flocculation, bouquet, and finish. Romanée-Conti is known for its perfumed aromas, perfect balance, silky texture, timeless aging potential, extraordinary quality, and unmatched flavor. Its burgundy has a deep reddish-brown hue, mild flocculation, and thin legs. The taste is fruity, ripe with tannins, high in acidity, and complex in flavor from red berries, black tea, and roses.[3]

To laypeople, sommeliers seem to possess a supernatural sense of taste. With a swirl and a sip, they can deduce the most delicate details in wine. Their idiosyncratic language is far from the normal appreciation of cocktails and beverages. But do they truly have a heightened sense of taste? Can we all, with a bit of practice, differentiate sweet, sour, and bitter flavors?

Not entirely. It turns out that some of us are born supertasters, while others have little or no sensitivity in the palate. Terms coined by Linda Bartoshuk, *supertasters* and *nontasters* are vastly different in their discrimination of taste.[4] While researching the genetics of this sense, Bartoshuk discovered that roughly 25 percent of people are extremely sensitive to the bitter compound 6-n-propylthiouracil (PROP). Meanwhile, an equivalent portion (25–30 percent) cannot taste it. The rest of the population is average in their ability to sense PROP. From these findings and others, we realize there is incredible diversity in taste among people.

Most individuals can appreciate numerous tastes while enjoying a plethora of cuisines. Each person has their favorite foods and preferred appetizers. Certain beverages are ordered again and again because of their distinctive taste. But generally, what average tasters eat and drink is flavorful enough that extra seasoning or sauces are unnecessary. We can tolerate spicy or bitter foods without a problem.

Supertasters have a superior taste, an inherited hypersensitivity. Eating is always an intense experience, no matter the occasion. Even the tiniest amount of bitterness can evoke powerful reactions. Sensitivity to sweet, salty, and savory dishes is also more pronounced. That's because supertasters have more papillae on their tongues compared with the average person. Their tongues are wholly covered in taste cells.

In contrast, nontasters are relatively insensitive to taste. Typical food is bland, requiring extra seasoning and sauces. They constantly add siracha and spices to liven up their dishes at mealtime. When it comes to dining choices, most are not overly picky. They have a much smaller number of taste buds, making solid likes and dislikes unusual. Nontasters seldom voice complaints but rarely give compliments.

While it may seem unfair to be labeled a nontaster when you can at least taste something, some individuals were born without taste at all—a condition called ageusia. Others lose their sense of taste entirely after injury or illness. Since the onset of the COVID-19 pandemic, millions of people have experienced taste insensitivity secondary to coronavirus infection. Many have faced severe, debilitating loss of taste (and smell) that has impacted their enjoyment of food. Of symptomatic patients reporting taste or smell insensitivity,

56 percent reported a decrease in their enjoyment of life.[5] What is loss of taste like? Adrian Wellock experienced taste insensitivity after contracting an upper respiratory infection. At first, he struggled with a lingering metallic taste in his mouth. Then one morning he realized his sense of taste was gone when a glass of orange juice tasted the same as water.

What is it like to eat and drink without a sense of taste? "I can feel textures, the viscosity of liquids, I know how chewy or light the food is, I can feel the spice (something you don't taste, you feel)," he said, "but I just can't extract any flavor from it. Bitter, sweet, salty, savory, or sour is irrelevant, as they all give the same experience."[6]

THE COMPLEXITY OF TASTE

Unsurprisingly, taste starts in the mouth, where there are somewhere between four thousand and ten thousand taste buds. These tiny sensory organs are located on the visible bumps on our tongues, otherwise known as papillae. Each taste bud contains as many as one hundred taste receptors. They send signals to the brain each time we eat or drink.

When you snack on an orange, the fruit is first broken down into smaller pieces through chewing. As your teeth grind down the flesh, it is mixed with saliva. This slightly alkaline secretion dissolves chemicals in the fruit, allowing them to enter the taste buds and interact with gustatory cells. Then the nerves running from the taste cells to the brain relay messages to the sensory cortex.

Citrus fruits taste both sweet and sour because of the combination of sugar and citric acid they contain. The more sugar, the sweeter; the more citric acid, the sourer.

Compared with oranges, grapefruits and lemons are sourer. But each person's taste buds have preferences about citrus fruits. A careful balance of sweet and sour makes each bite zesty and refreshing.

Most people can distinguish at least five tastes: sweet, sour, bitter, salty, and umami. Fat is the leading contender for a sixth taste.

The most recently identified of these, umami, is best described as savory or meaty. In the twentieth century, Japanese chemist Kikunae Ikeda studied the taste of tomatoes, meat, and cheese. With none of the four familiar tastes describing it accurately, he pinpointed glutamic acid as one of the core ingredients of each food. He called this unique taste *umami*, which means "good flavor" in Japanese.[7] Think of a mouthwatering steak sizzling on the grill: that's umami. As a seasoning, the salt of glutamic acid, monosodium glutamate (MSG), is often added to foods and soups.

Salty describes the taste of sodium ions when taste receptors bind to it—much easier to identify than umami. In the Amazon rainforest, the Yanomamo people ingest about two hundred to three hundred milligrams of salt daily. But the average American consumes 3,400 milligrams daily.[8] The right balance of salt is crucial to life and health. The body needs salt to maintain fluid balance, transmit nerve impulses, and move muscle fibers. Without salt, we will die. But too much leads to increased blood pressure and opportunities for stroke or heart attack.[9]

The most sensitive of tastes, bitter, has the widest variance in individual tolerance. Many bitter compounds are poisonous and toxic, which is why we usually perceive bitterness as unpleasant. The ability to differentiate bitterness is a protective mechanism that helps us avoid harm.

But bitterness in small amounts can balance other tastes and make food more enjoyable. Bitter melon, artichoke, and ginger are treasured by some but despised by others. Bitter hops are a favorite taste of many beer drinkers. Some coffee lovers can't last the morning without the bitter arousal of their favorite brew. In some dark chocolates, antioxidants add to a decadent and beloved bitterness.

Do you have a sweet tooth? Sweetness is often thought of as a pleasure-inducing taste. It is almost universally pleasant, signaling the presence of sugars such as sucrose and glucose. Unfortunately, sugar is so enticing to our taste buds that overindulgence is rampant. In the United States, it has led to the epidemic of obesity and other negative medical consequences.

When you taste something sour, it reflects the amount of acidity from organic food acids. High-acidity foods such as lemons and vinegar make our mouths pucker as the brain registers their taste. Small amounts of acidity can add flavor and zest to many foods. A squeeze of lemon juice can bring out the flavor in baked salmon. Vinegar added to salad dressing yields sour notes to the fresh ingredients. The fermented cabbage sauerkraut is a favorite German addition to meats and sandwiches. The Korean side dish kimchi is similarly fermented, albeit with different spices.

My favorite candies are Sour Patch Kids. As a child, I could suck on lemons all day, which explains why these sour gummies are addictive to me. What makes Sour Patch Kids so compelling compared with other candy? The answer lies in the advertising: "First they're sour. Then they're sweet." Most candy targets only a single taste receptor, but Sour Patch Kids are delicious because they challenge sweet and sour tastes.

Different combinations of taste add to its complexity, but a person's genetic makeup contributes to their perception of taste as well. People's sensitivity to sweet, sour, salty, bitter, and umami depends on their genes. For instance, variants of the gene TAS2R38 determine how individuals will react to bitter tastes.[10] Sensitivity to taste also affects dietary choices while impacting overall health. Individuals with a lower sensitivity to fatty taste are more prone to eat foods high in fat. Those who are repulsed by the taste tend to stay away.

Taste, smell, and flavor are all intimately linked. Sometimes the terms are used interchangeably. But each has a divinely designed meaning. As we learned in the previous chapter, every substance has a unique smell. We can sense the aroma of foods directly from the nose. But a large part of olfaction comes from the roof of the mouth as we eat and chew. This leads to some confusion because a portion of what we consider taste is actually smell.

What about flavor? Flavor is the sensory experience of food or liquid in the mouth. It is determined by both taste and smell. Of the two senses, the sense of smell is the dominant factor in determining flavor. While the gustatory system is limited to five (or six) main categories, the olfactory one can identify millions of different smells.

The flavor of a food or drink is easily changed by altering its smell while leaving the taste relatively unchanged. Flavors can also be added to change taste. If you visit the World of Coca-Cola museum, you'll be invited to sample different Coke products from all over the globe. Timeless classics whet your appetite, while exotic flavors from faraway places may seem strange. Though the taste of each beverage is similar, flavoring changes the impression of each drink.

Flavorists help food and beverage companies create unique flavors to attract our senses. Either organically available or artificially made, flavoring ingredients can enhance natural foods or add a kick to processed foods and drinks. Cinnamic aldehyde yields cinnamon. Limonene smells like orange. Methyl salicylate gives a wintergreen scent. Many others provide an abundance of flavors to enjoy when eating or drinking.

Taste, smell, and flavor contribute to our enjoyment of food and drink. The slightest alterations in taste and smell result in our favorite flavors of ice cream, tea, and soda. Too much or too little, and we cringe in disgust. Just the right amount, and we delight in ecstasy. In these moments, we experience the fullness of joy in what God intended for our bodies.

TASTE AND SEE

The motto "Eat, drink, and be merry, for tomorrow we may die" is often ascribed to Epicurus. According to his philosophy, pleasure is the chief goal in life. Epicureans strove to live to the fullest while obtaining the most significant amount of pleasure in all things. These include diversions of the mind and gratifications of the body, though philosophical dialogue is more highly esteemed than desires of the flesh.

Throughout the Bible, similar words can be found, though their meaning is subtly different. Solomon tells us that "everyone should eat and drink and take pleasure in all his toil—this is God's gift to man" (Eccl. 3:13). From breakfast to dinner, we recognize that our daily sustenance is a gift from God. Our enjoyment comes not from a hedonistic philosophy but from a heart of gratitude that reflects thankfulness.

Ecclesiastes speaks of *hebel*, translated as "meaningless" or "vanity." This Hebrew word is central to the book's theme. The writer is searching for meaning. After looking far and wide, he discovers that life is meaningless, full of vanity and futility. Our lives are a mystery that we are incapable of understanding. No matter how hard we try, how diligently we work, or how loving we are, everything is a chasing after the wind.

The only solution to this dilemma is to look to the heavens. Yahweh is the author of our lives. He has made everything beautiful in its time, putting eternity in our hearts. Yet we still cannot fathom what God has done from the beginning to the end (Eccl. 3:11). Though we may never understand, true freedom lies in choosing to "fear God and keep his commandments" (Eccl. 12:13).

When we trust in God's sovereignty, our worldview is transformed. Everything is beautiful and nothing is meaningless because "all things work together for good, for those who are called according to his purpose" (Rom. 8:28). Even amid broken relationships, untreatable disease, or financial catastrophe, we can remain joyful in our affliction. Doing so doesn't downplay our hurt and pain but reminds us that ultimately God is good.

The psalmist says, "Taste and see that the Lord is good!" (Ps. 34:8). How interesting that the songwriter chooses a taste metaphor. Savor God. Let all your senses, including taste, relate to the wonder that he is good.

When we believe that God is good, we more readily enjoy the life he has ordained for us. Though we may never have all the answers to complex questions, we can still find abundant joy.

How? If there is food before you, don't just eat it for nutrition. Take time to enjoy the taste, aroma, and flavor. From the spicy, sweet, and savory taste of coconut in massaman curry to finger-licking-good piri-piri chicken from Mozambique, food is a heavenly gift. When eating Portuguese pastel de nata, enjoy every bite of the rich, flaky pastry with its soft, mouthwatering custard.

The ability to savor is a gift to us, and our enjoyment is consciously or subconsciously a tribute to the Creator. Savoring is in essence an incarnational experience when we recognize that both the flavors and the ability to appreciate them are from God's hand.

If your favorite beverage is on tap, savor its unique flavor. During the winter, allow the rich and aromatic decadence of hot chocolate and marshmallows to awaken your taste buds while warming your soul. In the summer, cool down with a refreshing Thai iced tea, a favorite with spicy Southeast Asian dishes. The addition of sweetened coconut milk to traditional Thai black tea is sure to stimulate your senses.

When we take time to taste and see, we realize God is good. It is he who created our taste buds to sample creation. His genius allows flavors to explode in our mouths, both organic and artificial. The perfect balance of taste and smell tickles our senses and captivates our hearts.

Despite the theme of *hebel* in Ecclesiastes, everything is meaningful when viewed, smelled, and tasted through divine perspective. There is significance in simple foods and modest wine. We can eat our bread with joy and drink our wine with merry hearts, for God approves of what we do (Eccl. 9:7). That's why the writer of Proverbs exhorts us to "eat honey, for it is good, and the drippings of the honeycomb are sweet to your taste" (Prov. 24:13).

In 1 Samuel 14, Saul instructs his men not to eat food until he avenges his enemies. As they come to the forest, honey lies on the ground, but the men refrain from eating. Jonathan, Saul's son, has not heard of his father's command. So he dips his staff in honeycomb and savors the flavor. Immediately, his eyes become bright from the taste. Not only was the honey sweet but its sugar strengthened his countenance.

Foods delight our senses, yet they are also a needed source of energy. During their exile, the Israelites complained of hunger, fearing they would starve to death in the desert (Exodus 16). Although they were free, they longed to return to captivity. At least then they had pots of meat and ate all the food they wanted as enslaved people (although it's hard to imagine that the enslaved would ever have "all they wanted" to eat in a culture where even the supply of straw to make bricks was rationed).

When God heard their grumbling, he provided manna from heaven in the morning and quail in the evening. Described as thin flakes covering the ground like frost, manna was ground up to make bread and cakes. White in color, like coriander seed, it was compared to wafers made with honey (Exodus 16; Numbers 11).

Even in taste, manna is a symbol of God's goodness. As the bread of angels (Ps. 78:25), it foreshadows Jesus, the ultimate bread of life. Both are gifts from heaven, but while manna provides physical sustenance, Christ offers eternal life. Just as manna was a reminder of God's faithfulness to the Israelites every morning, today communion commemorates Jesus' sacrifice each time we partake.

Jewish literature ascribes a unique taste to manna, adapted to each individual's palate. For adults, it was hearty and full of substance. To infants, it tasted like breast milk. The Talmud compares manna to the many tastes of breast-feeding, where breast milk takes on the flavor of the most recent meal of the mother. Its adaptable nature allowed it to be ground in hand mills, beat in mortars, and boiled in pots (Num. 11:8).

Today, Italian farmers in Calabria and Sicily cut off the sap of the tamarisk tree to obtain what they call manna. Hedysarum manna and Shir-Khesht manna are imported from Iran. Chef Paul Liebrandt used the latter in a dish with kindai kampachi, fresh wasabi, and charred apricots. According to him, the texture is uniquely chewy and crunchy: "It makes the food intensely personal because no two people taste manna the same way. I might taste a haunting minty-ness, while you might detect a whiff of lemon. No other ingredient is like that."[11]

That flavor profile has God's touch all over it.

THE CONNECTEDNESS OF CREATION

During creation, God said to Adam, "Be fruitful and multiply and fill the earth and subdue it, and have dominion over the fish of the sea and over the birds of the heavens and over every living thing that moves on the earth. . . . Behold, I have given you every plant yielding seed that is on the face of all the earth, and every tree with seed in its fruit. You shall have them for food" (Gen. 1:28–29). On the third day, God made seed-yielding plants, fruits, and vegetables. Two days later, marine animals filled the seas while birds flew through the sky. Then on day six, livestock and wild animals began roaming the earth. After creating man and woman in his image, God gave them dominion over all living creatures and plants.

Green, leafy vegetables are known for their minerals, vitamins, and fiber. Low in calories, they reduce the risk of heart disease, high blood pressure, and obesity while improving mental health.

Good for the bones, peppery-tasting arugula is packed with vitamins A, C, and K, calcium, and manganese. Delicious when stir-fried with garlic, bok choy contributes to gut health with antioxidants and magnesium. Rich in iron and carotene, juicy and flavorful spinach is an excellent ingredient in soups, salads, and sandwiches.

"An apple a day keeps the doctor away" is a timeless saying. Fruits are among the most nutritious, healthy, and delicious foods. Globally, thousands of different fruits are available to whet your appetite. Consistent variety is a key to providing lasting benefits.

Blueberries' sweet but tart taste makes them a quintessential snack for children. A common ingredient of homemade pancakes, muffins, and other delicacies, their anti-inflammatory and antioxidant properties lower the risk for diabetes, heart disease, and certain types of cancer. Known as the king of fruits, mangoes have a tropical, floral taste. With high nutritional value, they are packed with vitamins and minerals, making them great for the skin, eyes, heart, and digestive system.

The wonder and delight of flavor couple with the provision for our bodies' needs. According to Genesis 9:3, every moving thing that lives is food for humanity. But for centuries, Jews ate only certain meats. Leviticus 11 and

Deuteronomy 14 prohibited unclean foods from being consumed. These included animals without cloven hoofs and those that did not chew their cud, shellfish, creeping creatures, and fish without fins and scales.

The life, death, and resurrection of Jesus Christ transformed the law. Everything became permissible, though not always beneficial (1 Cor. 10:23). Nowhere is this more evident than in Acts 10:9–16. As a devout Jew, Peter did not consume the unclean foods specified by the law. Yet one day God gave him a spectacular vision.

Imagine a gigantic sheet descending from the heavens. As it opens, all kinds of wild animals, sea creatures, birds, and even reptiles appear in it. Then God says to Peter, "Kill and eat." What was Peter's response? He was appalled! How could he possibly eat unclean food? Never once had he wavered in his observance of the law. Why would he start now? So he answered, "By no means, Lord."

Then God provided one of the most remarkable reversals of the new covenant. He turned the Jewish law upside down by declaring, "What God has made clean, do not call unclean."

We all know Peter as stubborn, brash, and hardheaded. The Lord needed to repeat himself three times before the sheet was finally taken up to heaven. In the end, the message finally pierced Peter's thick skull. He traveled to the home of Cornelius, where he shared in a feast of once-forbidden food while introducing the gentiles to the good news of the kingdom.

In Romans 14, Paul teaches us that nothing is unclean in itself but is unclean only for the one who thinks so. Some believe they can eat anything, but others will partake only of vegetables. Paul rebukes those who pass judgment on others—either those eating meat or vegetarians. Ultimately, each person's conscience dictates what they eat, "for whatever does not proceed from faith is sin" (Rom. 14:23). Whether a vegetarian, flexitarian, pescatarian, vegan, or meat lover, we are free to enjoy the taste of our preferred foods.

Regardless of our convictions and preferences in flavor and aroma, we must remember not to judge others. Not only should we not condemn but we should never be a stumbling block or hindrance to other brothers and sisters (Rom. 14:13). Whatever they consume, we should respect their tastes and honor their choices.

As we choose what to eat and drink, our dedication to Jesus helps us remember that all creation is interconnected. Though we have dominion over all the earth, the choices we make have a profound impact on everything and everyone around us. The mandate of Genesis 1 reminds us to care for God's creation while stewarding its resources. The lands and seas, vegetation, and living creatures provide pleasure in this life and blessings for generations to come.

Food, wine, and beverages were created for our enjoyment. In 1 Timothy 4:4, Paul teaches us that "everything created by God is good, and nothing is to be rejected if it is received with thanksgiving." But even if we consume nourishment with thanksgiving, we must be wary of overindulgence. In Ephesians 5:18, Paul contrasts the danger of alcohol intoxication with the filling of the Spirit.

There is a fine line between the sin of gluttony and the enjoyment of taste, smell, and flavor. The physical effects are easier to observe: gluttony can lead to feeling tired, sluggish, drowsy. But the spiritual consequences are far more subtle. Overindulgence can twist the goodness of

God's creation into idols that have authority over our lives. Such idols not only affect our bodies and acts of worship but also poison our relationships with those we love. For those struggling with alcohol addiction, what was meant to be a blessing in the context of fellowship is a curse that steals life away.

FOOD, FELLOWSHIP, AND THE BREAD OF LIFE

Every fall at our church, small groups begin anew. In the year to come, brothers and sisters will fellowship at someone's home to explore issues of faith. The truth of the gospel will be shared. A genuine openness to newcomers will be encouraged, with lively discussion often the norm.

One essential ingredient is the hospitality of food and drink served during each gathering. Whether it's shepherd's pie as an entree, Caesar salad with garlic bread as an appetizer, or a dessert of snickerdoodle cookies, small-group members always enjoy the taste of food and the company of friends.

In Acts 2:42, Paul describes believers as devoted "to the apostles' teaching and the fellowship, to the breaking of bread and the prayers." The early Christian church was marked by the Greek word *koinonia*, defined as "having in common." Fellowship meant participation by giving and receiving one's blessings with others. One of the greatest gifts was sharing tasty food and flavorful drink.

For the early church, the breaking of bread and the sharing of meals was an integral part of the life of the people. Christians loved sharing meals. Eating together demonstrated their love for one another and desire to serve one another. They ate together often.

Research from the University of Oxford reveals that the more often people eat with one another, the happier and more satisfied they are with their lives.[12] The combination of food and fellowship enhances one's sense of connectedness and contentment. Communal meals increase social interactions, resulting in more friendships.

From the days of the Torah, feasts and festivals marked the Jewish calendar. These gatherings were celebratory, commemorative, and communal. Drawn together by them, the Israelites worshiped in community. Remembering the work of the Lord, they kept alive the story of God's redemption. Each time, food was symbolic, nourishing, and delightful.

Every Thanksgiving, we remember the blessings of the harvest and the provision of God. The hallmark of Thanksgiving is giving thanks to the Lord while sharing laughter and conversation over a hearty meal. Food and wine taste sweeter in the company of family and friends. Certain holiday foods provide even more reason to celebrate.

You can partially thank the neurotransmitter serotonin for the unifying, celebratory power of food. It plays several critical roles in the body. It affects memory, learning, and happiness while regulating sleep, hunger, and sexual behavior. Known as the feel-good chemical, it influences emotional stability and focus. Low levels have been linked to depression. That's why many antidepressant medications increase serotonin levels as treatment.

Foods that augment serotonin production include turkey, spinach, and bananas. Turkey and bananas contain tryptophan, the amino acid that is converted into serotonin. Spinach is high in folate, the vitamin used to help create serotonin. When these foods are served, it's no wonder a synergistic effect results in good cheer and glad tidings. We can't help but feel calm and more relaxed. Add a little wine, and everyone is sure to be pleased.

In Acts 2, fellowship and the breaking of bread contributed to the early church's exponential growth. But the most important ingredient was the Bread of Life. In John 6:35, Jesus says, "I am the bread of life; whoever comes to me shall not hunger, and whoever believes in me shall never thirst." For early Christians, koinonia, prayer, and the sharing of food centered around the apostles' teaching about Jesus.

Quoting Deuteronomy 8:3, Jesus says, "Man shall not live by bread alone, but by every word that comes from the mouth of God" (Matt. 4:4). The sweet flavor of manna was a delight to the Israelites as they wandered the desert. But the Word of God is the only sustenance for our souls. No matter how delicious the food we eat or mouthwatering the liquids we consume, nothing can compare to the living Word.

When we embrace our sense of taste the way God designed it, we experience the fullness of joy while inviting others to do so as well. "How sweet are your words to my taste, sweeter than honey to my mouth!" (Ps. 119:103).

QUESTIONS FOR REFLECTION

1. Do you consider yourself a supertaster, a nontaster, or normal in your sense of taste?
2. Have you experienced taste insensitivity? If so, how did it affect you?
3. The psalmist says, "Taste and see that the Lord is good!"

(Ps. 34:8). What does this verse mean about God? Can you think of other Scriptures that speak to taste?

4. How does overindulgence in food, wine, and drink affect you physically and spiritually?
5. Which Bible verses speak to the relationship between nourishment and fellowship? How do the two interrelate?

NOTES

1. Book, T. (2022). Romanee Conti 1945. Retrieved August 7, 2022, from https://tastingbook.com/wine/domaine_de_la_romaneeconti/romanee_conti_1945?language=en
2. Crum, G. (2018). *Le Domaine de la Romanée-Conti.* Lannoo.
3. Crum, G. (2018). *Le Domaine de la Romanée-Conti.* Lannoo.
4. Bartoshuk, L. M. et al. (1994). PTC/PROP tasting: Anatomy, psychophysics, and sex effects. *Physiol Behav* 56(6), 1165–71.
5. Coehlo, D. H. et al. (2021). Quality of life and safety impact of COVID-19 associated smell and taste disturbances. *American Journal of Otolaryngology* 42(4).
6. Wellock, A. (n.d.). Adrian's story: Living without taste—A story of ageusia. Fifth Sense. Retrieved August 7, 2022, from https://www.fifthsense.org.uk/stories/adrian-wellocks-story
7. Ikeda, K. (2002). New seasonings. *Chem Senses* 27(9), 847–49.
8. Oliver, W. J. et al. (1975). Blood pressure, sodium intake, and sodium related hormones in the Yanomamo Indians, a "no-salt" culture. *Circulation* 52(1), 146–51.

9. Appel, L. J. et al. (1997, April 7). A clinical trial of the effects of dietary patterns on blood pressure. DASH Collaborative Research Group. *N Engl J Med 336*(16), 1117–24.
10. Prescott, J., & Tepper, B. J. (2004). *Genetic variation in taste sensitivity*. Marcel Dekker.
11. Arnold, D. (2010, June 8). Ancient manna on modern menus. *New York Times*. https://www.nytimes.com/2010/06/09/dining/09manna.html
12. Dunbar, R. I. M. (2017). Breaking bread: The functions of social eating. *Adapt Human Behav Physiol 3*(3), 198–211.

5

HEAR

After his family moved from another state, Bobby, my new patient, came in for his eighteen-month doctor's visit. As I introduced myself, he smiled but didn't say anything. Assuming he was shy, I began the physical examination while conversing with his mother. Bobby was growing well, at the eighty-fifth percentile for height and weight.

According to his mom, Bobby was a healthy boy. Though he missed several well-child visits because of their move, he didn't have any pressing medical issues. Her only concern was his worsening behavior. At home, Bobby was often defiant and prone to temper tantrums. Compared with other children, he seemed to whine a lot. Sometimes his unpleasant behavior progressed to full-blown meltdowns. Recently he'd started hitting and kicking others, sometimes even biting.

Her description seemed odd for the boy who was patiently sitting in front of me with a sly grin. Was it simply the front edge of the terrible twos, when kids begin developing independence and a sense of self? At that age, toddlers explore their environments on their own terms. But because their emotional and verbal skills are just starting to develop, they become easily frustrated.

As I pondered whether his behavior was normal, I explored other physical and psychological possibilities. When I asked how many words he spoke, I was surprised when his mom said he "hardly speaks any." This led me to question how he responded to her voice. Did he understand what she was saying?

His mother turned pale, staring out the window as if she'd seen a ghost. With a sudden epiphany, she realized Bobby had never responded to her spoken words. Whenever his back faced her, it was as if he was ignoring her. Thinking back to his infancy, she realized noises did not startle Bobby. He reacted only when he could see someone talking.

After sensing what was wrong, I ordered several hearing tests. Each confirmed what I suspected: Bobby was deaf.

Though 98 percent of newborns are screened for hearing loss every year, some fall through the cracks. Roughly three in one thousand infants in the United States are hard of hearing. In 2019, six thousand babies were identified as having permanent hearing loss.

Helen Keller once said, "Blindness cuts us off from things, but deafness cuts us off from people."[1] Born in 1880, Keller lost her vision and hearing at nineteen months. It's likely bacterial meningitis was the culprit.[2] Though it left her blind and deaf, she still graduated from Radcliffe College and paved the way for others to follow. As a prominent writer, she became a staunch advocate for the disabled.

The absence of sound affected Helen more profoundly than the inability to see. During her lifetime, few treatments existed for either disability. Though she could use both senses for only a short while after birth, she longed to hear the voices of those near and dear. She once shared, "The sound of the voice brings language, sets thoughts astir and keeps us in the intellectual company of man."[3]

Hearing loss affects a person's ability to develop speech and language. Children with auditory processing difficulties are at high risk for developmental delay. It impacts their communication, learning, and social skills.

For young children like Bobby, vocabulary acquisition is stunted.[4] Concrete words such as *cat*, *run*, and *two* are more easily learned than abstract words such as *before* and *after* or function words like *an*, *are*, and *a*. Expressions with multiple meanings such as *bank*, *swing*, and *park* are challenging to comprehend. This makes complex sentences hard to understand.

When learning to speak, children with hearing loss have greater difficulty with quiet sounds like "s," "sh," "f," "t," and "k." They tend to leave them out of speech. Depending on their level of deafness, they tend to shout or whisper. Without appropriate intervention, their academic achievement may be severely limited.

Not surprisingly, children with hearing loss often feel alienated and alone. This stems from a limited ability to socialize with other children their age. When playing in group settings, they have a hard time making friends, which can lead to increasing frustration and despair.

Today, implants, hearing aids, and speech therapy offer newfound hope for the deaf and hearing impaired. Early intervention and identification of hearing loss are crucial for a child's behavior and development. They can halt the progression of development delay while helping foster language acquisition and social skills.

Hearing aids, both external and bone anchored, intensify sounds.[5] Infants, children, and adults of all ages can benefit from them. Before babies can speak, these devices enhance their ability to hear sounds and learn. For many elderly people, hearing aids are a great way to circumvent deafness when hearing loss begins.

Cochlear implants can be used when hearing aids are inadequate for those with more severe hearing loss. While hearing aids amplify sounds, cochlear implants bypass damaged parts of the ear. They send signals directly to the brain by stimulating the auditory nerve. Even if the auditory nerve is injured, brain-stem implants can provide sound. These devices bypass the inner ear and auditory nerve, connecting directly to the brain stem. People born deaf now have an array of technological devices at their disposal. Unlike for their predecessors, there is hope for them to hear loud and clear.

PRECISION HEARING

During his earthly ministry, Jesus often spoke to large crowds. Several times he said, "He who has ears to hear, let him hear." Though he was encouraging the people to pay attention to his words, those farthest away from him likely could not make his words out clearly. During those days, there were no megaphones or portable speaker systems. People relied solely on their auditory organs to hear.

Several years ago, I was invited to speak at a large conference. The Holy Spirit stirred the people's hearts as the praise team led everyone into the heart of worship. During the final praise song, I remembered a typographical error on one of my sermon slides. So I subtly moved toward the audio-visual team in the back of the room.

As I communicated my mistake to the person in charge, he could hardly hear me through the background music. Raising my voice didn't seem to help at all. Then he cupped his hands around his ear while directing it toward me. Miraculously, he was able to make out my words and quickly changed the slide just in time for the sermon. You've probably done something similar before.

The visible part of the outer ear, the pinna, has a characteristic shape and curve. It captures, amplifies, and directs sound waves toward the ear canal. Cupping our hands behind our ears while pulling them forward funnels sound waves into the ear. In a crowded room, we're better able to hear others by cupping our ears toward them.

The larger the pinna, the more sensitive the ear is to sound. An elephant's enlarged auricles allow them to hear from as far as 100 to 150 miles away.[6] Bat-eared foxes have unique pinnae that enable them to listen to the movements of insects thirty centimeters underground.

The unique twists, folds, and ridges of human pinnae help us hear the voices of friends, family, and coworkers. Our ears' design amplifies the volume of sound waves in the 2,000 to 3,000 hertz range. This is the frequency where many consonants reside in our language. Because our ears enhance sounds fifty to one hundred times, we can better interpret words when others speak.

Sound waves travel from the pinna into the ear canal. As they hit the eardrum, the sound waves cause three small middle-ear bones to vibrate. The malleus, incus, and stapes are the tiniest bones in the body. Collectively, they are known as the auditory ossicles (Latin for "tiny bones").

The drumming motion of the middle-ear ossicles causes fluid in the inner ear to shift. The inner ear is comprised of two main parts: the cochlea and the semicircular canals. The former is responsible for hearing, the latter for balance. *Cochlea* comes from the Greek word for "snail," *kohlias*, because of its characteristic shape.

The organ of Corti is located inside the cochlea. The body's microphone, it contains a small array of sensory cells, specific areas that recognize particular frequencies of movement. The organ of Corti is surrounded by two membranes, one of which is the basilar membrane. The shifting fluid in the cochlea causes hair cells on the basilar membrane to bend. Hair cells convert the fluid vibrations into electrical impulses. These impulses are transmitted to the auditory nerve, which carries the signals to the brain.

We are born with roughly sixteen thousand hair cells in the ear. Two types exist. About twelve thousand amplify sound. The remainder generate electrical impulses relayed to the brain. Together, they help us detect even a single air molecule when it hits the eardrum. Over time, they can be damaged or destroyed, significantly affecting our ability to hear.[7]

A range of 20 to 20,000 hertz is distinguishable for humans. But frequencies between 250 and 6,000 hertz are the most important, lying in the audible range of speech. Vowel sounds make up the lower frequencies, while consonants like "f," "s," and "th" contribute to higher frequencies.[8]

Pitch is the frequency of sound waves processed by the ears and interpreted by the auditory cortex. Music incorporates pitch by tying sound to letters in the alphabet. For musicians and vocalists, middle C has an exact frequency of 262 hertz, while the first A above it vibrates with a frequency of 440 hertz.

Our praise and worship team always tunes their instruments before practice. Guitars, pianos, and basses go out of tune because of frequent use, changes in humidity and temperature, and damage. This results in notes being flat (too low) or sharp (too high). For symphony orchestras and concert bands, instruments are tuned to the 440 hertz A. This allows everyone to play harmoniously.

About one in ten thousand people is born with perfect pitch. Though I never inherited this gift, my father did. As a choir director, he can tell when vocalists are flat or sharp, when the sound waves they're creating are slightly below or above the standardized frequency for each note. Ask him to sing middle C, and you will hear the exact note at 262 hertz. Play him any note of your choosing, and he will identify it perfectly.

Though many of us will never experience perfect pitch, we have the propensity to learn relative pitch. This is the ability to identify musical notes in relation to others. With practice and study, playing by ear is also possible—not by reading notes but by duplicating what was once heard. This remarkable plasticity of hearing speaks to the capacity of the auditory system to adapt and acquire new skills. God has given each of us an innate ability to learn. According to Romans 10:17, "Faith comes from hearing, and hearing through the word of Christ." By nurturing our auditory gifts, not only do we develop relative pitch but we learn to hear the voice of God.

THE VOICE OF GOD

At thirteen, I attended a Jewish friend's bar mitzvah. It was a joyous occasion with an elaborate feast following the ceremony. Though outwardly extravagant, the bar mitzvah signified the coming of age for this adolescent. He was now a full-fledged member of the Jewish community, with all the privileges and responsibilities that came with it.

In Hebrew, *bar mitzvah* means "son of commandment." As one subject to the law, my friend was now obligated to observe the Torah, the first five books of the Old Testament. He studied diligently, memorizing significant parts of the law in the weeks and years prior to that day. At the ceremony, he read from the written Torah while also giving an oratory discourse on the law.

Since the time of the patriarchs, the oral Torah was passed down from generation to generation before finally being written down. For Jews, listening to the law was vital to committing it to memory. They listened to every statute, law, and legal interpretation with keen ears. Not only did the Jews hear the law but also they were subject to it.

In the Old Testament, God appears as a burning bush, a cloud by day, and a pillar of fire by night. While revealing himself, he speaks to Adam and Eve, Cain, Noah, Abraham and Sarah, and Moses. Though not always seen, God is heard by those he speaks to.

Sometimes God's followers responded with an abundance of faith. On other occasions, they disregarded his voice. In Genesis 2:17, God instructs Adam not to eat fruit from the tree of the knowledge of good and evil. Yet Adam and Eve cannot resist the allure of that fruit. Their cardinal sin leads to the fall of humanity.

In Exodus 3 and 33, God speaks directly to Moses. Having seen the misery of his people in Egypt, he calls Moses to deliver the Israelites from Pharaoh. At first, Moses is reluctant to obey. He offers various excuses, vehemently protesting his calling.

As the Lord's anger rises, Moses finally responds in faith. He leads the Israelites from exile to freedom. Though he never entered the promised land, it is extraordinary to witness the transformation of his relationship with God. Early on, Moses is distant and afraid to encounter the living God. But in his final years, he listens to God's voice as if speaking to a friend (Ex. 33:11).

Similarly, Abraham responds in faith each time he hears God's voice. When the Lord calls him to take Isaac to the land of Moriah and sacrifice him as a burnt offering, he does so without hesitation. Abraham carries forth the command that most of us would never think of doing in faith. Because of this, the Lord blesses him and promises him offspring as numerous as the stars in heaven and the sand on the seashore (Gen. 22:17).

Throughout the Scriptures, hearing is instrumental to faith. Only when we hear the voice of God can our lives be transformed. John compares the people of God to sheep who recognize their shepherd when he speaks (John 10:27). For him, "Whoever is of God hears the words of God" (John 8:47).

Yet when God speaks, we don't always recognize his voice. Even the prophet Samuel heard God's words as a boy and yet did not recognize him. When the Lord calls him, Samuel runs to the priest Eli and says, "Here I am, for you called me." Yet it is not Eli who calls. God speaks to Samuel three times before Eli instructs Samuel that God is calling him (1 Sam. 3:1–10).

How can we differentiate God's voice from the world's chatter? The first step is pursuing God with our whole hearts. The Lord shares this simple truth in Jeremiah 29:13: "You will seek me and find me, when you seek me with all your heart." Only the greatest of devotion allows us to hear and understand the longings of loved ones.

Second, we must renew our minds. In Romans 12:2, Paul urges us, "Do not be conformed to this world, but be transformed by the renewal of your mind, that by testing you may discern what is the will of God, what is good and acceptable and perfect." Sometimes God speaks through audible words that vibrate the tympanic membranes and hair cells in our inner ears. At other times, he communicates directly to our minds through the Holy Spirit. When we constantly renew our minds with the Word of God, we develop the psyche of Christ.

Finally, we need to set aside time to listen to God. "For the eyes of the Lord are on the righteous, and his ears are open to their prayer" (1 Peter 3:12). By devoting time to prayer, we prepare our senses to receive the Lord's communication. Listening leads to the expectation of response. In those moments, we become friends of God, like Moses.

DESTRUCTIVE NOISE

At seventeen, I attended my first rock concert. It was exhilarating. I could feel the crowd's energy as I entered the indoor venue. The guitarist played the first riff, and the stage vibrated as the people roared. Then the lead singer sang the first stanza, each word mesmerizing the fans. His high-pitched voice reverberated through the walls. I still remember the drummer beating away at the cymbals as the loudspeakers pulsated.

The electrifying music played on repeat in my head that

night. But there was also a faint ringing in my ears long after the concert was over. I was grateful the buzzing was gone after a few days. I wish I had known that day what I learned much later. For some people, the noise never fades.

The greatest threat to our hearing is sound itself. Indoor rock concerts range from ninety to 120 decibels in sound. At 120 decibels, ear pain can occur, with hearing loss possible in less than two minutes.[9] The damage is permanent.

At first glance, sixteen thousand hair cells seem like plenty for any person. But compared with the millions of optic nerve cells we're born with, our auditory sensory cells are in short supply. They need to last a lifetime. Almost everyone has far fewer hair cells than they were born with. In the United States, one in four adults has signs and symptoms of hearing loss.[10]

The degradation of hair cells reminds me of the annual wildebeest migration in the Serengeti. Every year, giant herds of grazers migrate across northern Tanzania and Kenya in search of green pastures. Before they arrive, the grazing lands are lush with grass. When individual animals run across the fields, the blades of grass, after being stepped on, immediately rebound to their standing position. But when a migration of wildebeests repeatedly tramples the grasslands, the grass cannot recover.

Auditory hair cells are similar to the blades of grass. They bend back and forth as the fluid moves around in the cochlea with sound vibrations. Many people complain after a concert or sports game that they can't hear as well as before. But after a short period, the ears recover as hair cells bounce back and heal. But if sounds are too intense, frequent, or loud, hair cells are damaged permanently. The problem is, lasting changes are insidious. Up to 40 to 50 percent of hair cells can be destroyed before noticeable changes can be appreciated on a hearing test.

It's intuitive for individuals to keep their iPhone AirPods at reasonable levels. Most people realize when sound is too loud for their ears. At 130 decibels, airline jets are noisy during takeoff.[11] To protect our hearing, we cover our ears, and airport workers wear ear protection on the tarmac. But instantaneous loud sounds like gunshots can cause immediate damage without warning.

Repeated exposure over time explains why age is one of the most significant predictors of hearing loss. Every decade, our hearing declines. Sixty-eight percent of baby boomers experience hearing loss, compared with only 7 percent of millennials.[12]

In the gospel of Mark, Jesus warns us to "pay attention to what you hear: with the measure you use, it will be measured to you, and still more will be added to you" (Mark 4:24). Not only do the quality and quantity of sounds affect us but also the content of what we hear and process. Loud sounds cause physical destruction to our auditory systems. Yet the more significant concern is the damage to our spiritual lives from the world's commotion. In an era of social media and online influence, innumerable voices vie for our attention and emotions while clouding our minds. It's hard to sift through all the clatter when false teaching is rampant.

Proverbs 12:18 warns us about the destructive voices that surround us. They are devastating and deadly, piercing like a sword. In the past, the sources of these influences were limited. Today, they arise from countless places. The internet, livestreaming, and instant messaging applications bring us into constant contact with potentially unhealthy noise.

In some countries, 46 percent of Gen Z get their spirituality from TikTok.[13] Take social-media influencers, for example. They are paid to look beautiful. Their clothes are hip and trendy. But the more we see their ads and hear their voices, the more troubled we feel about our flaws and imperfections. They can make us think we're overweight, ugly, and poor. From perfectly sculpted bodies to lavish homes, the appearance and lifestyles and habits of others can easily become the object of our desires. These kinds of damaging words can make their way into our inner monologues, the voices in our heads, and their presence is deafening. Just as with our auditory hair cells, repeated exposure leads to irreparable consequences. In contrast, according to the writer of Proverbs, healing comes from the words of the wise.

The voices of the world come in a variety of forms. Some are loud and obnoxious, like the roaring lion on the prowl, ready to devour (1 Peter 5:8). At other times, we must beware of false prophets who come to us in sheep's clothing but inwardly are ravenous wolves (Matt. 7:15). Subtle shifts in theology may seem harmless at first glance. In time they rob us of our faith as the Word becomes diluted. If we are not careful, orthodox belief unravels into sects and cults.

Protecting our hearts and minds from worldly noises is similar to protecting our ears from deafness. Frequent contact with unhealthy voices leads to normalization and assimilation. For addictive behaviors, dependence becomes surreptitious. But just like sounds above 120 decibels, a single exposure to a false claim or wicked lie can also be damaging.

Destructive voices cripple our faith. They deceive us into thinking we're unworthy of God's love. They tell us we don't need a savior. They make us second-guess our thoughts and decisions, all deceits emanating from the father of lies. Sometimes these voices come from others, but on other occasions they arise from ourselves. Both are lies and we must guard against unhealthy external and internal voices.

Only the truth of the Word can remedy doctrinal falsehoods. We condition our minds to defend against heresy by consistently hearing the truth. When we speak on behalf of the gospel, we protect the integrity of the Christian faith. The more we do so, the more confidence we develop.

When the body of Christ collectively communicates words of encouragement, we are a witness to the world. Paul encourages us to let come from our mouths not unwholesome talk but only what helps build others up according to their needs (Eph. 4:29). These expressions of love encourage all who hear. By sharing them, our voices provide healing instead of death.

REDEEMING OUR EARS

As I entered the pediatric ICU to check on patients one morning, I noticed a teenage girl who was newly admitted. She was on a ventilator, a machine helping her breathe while she recovered from a traumatic brain injury. There were no other fractures or obvious puncture wounds on her body. As I reviewed her chart, I wondered what had happened.

The chart revealed an epidural hematoma, a collection of blood between the skull and the membrane covering the brain. It is usually accompanied by a skull fracture, which she also suffered. During the night, the neurosurgeons performed an emergency craniotomy and evacuated the blood.

The prognosis was good, given her youth and response

to the surgical treatment. But what was the cause? Car surfing.[14] I had not heard of the phrase, but her mother was very familiar with the term. This girl had been riding on the outside of a moving vehicle. Most commonly, teenagers sit on the roof of a car or in the back of a pickup truck.

According to the mom, her daughter and her daughter's friends were addicted to car surfing. Someone would drive their vehicle each weekend while everyone else would stand or sit on top of it. They knew it was dangerous, but that was part of the appeal. They recorded the events with their smartphones while sharing each daring feat on social media.

The mother repeatedly warned her daughter of the risk of severe injury. But she would not listen. Her parents were forced to ground her and take away her phone. The night before, the defiant teenager sneaked out of the house to join her friends.

Teenagers are receptive to peer pressure, risky behavior, and short-term pleasure. Sometimes they hear, but often they don't listen. Perhaps this innate defiance led God to decree to Moses that "anyone who curses his father or his mother shall surely be put to death" (Lev. 20:9). While this statement may be hyperbolic, it speaks to the rebellious nature of the heart.

Though we may be quick to dismiss car surfing or subway surfing as activities we would never engage in, each of us struggles with peer pressure to some extent. Whether it's needing to own the latest gadgets, wanting to live in the most desirable parts of town, or indulging in unhealthy foods, everyone suffers from thorns in the flesh.

In Romans 2:13, Paul says, "For it is not the hearers of the law who are righteous before God, but the doers of the law who will be justified." It's easy to listen to someone but much harder to heed their words. It isn't only of teens that this is true. Even when we identify wise counsel, it's often difficult to accept. The girl with the epidural hematoma knew the dangers of car surfing, yet she couldn't resist the temptation.

For Christians, the redemption of our hearing lies not only in listening to the words of God but also in following his voice. John tells us that "whoever is of God hears the words of God" (John 8:47). Hearing is not enough, but it is requisite for faith.

The oral Torah was earmarked for the Jews. But the gospel is for all men and women. According to Hebrews 4:2, the good news came to Christians and Jews but the message they heard did not benefit the latter because they were not united in faith with those who believed.

When we listen to Jesus' words and submit to his instruction, our lives are redeemed for the kingdom. Jesus says, "Everyone then who hears these words of mine and does them will be like a wise man who built his house on the rock" (Matt. 7:24). When the storms of life come, our foundation is solid. But those who hear Jesus' words and do not heed his counsel will be like the foolish man who built his house on sand.

The book of James tells us we deceive ourselves when we are hearers of the Word but not doers of it (James 1:22–25). James uses the metaphor of a person who looks intently at their face in a mirror, then immediately forgets their appearance. So is the person who hears and does not do. But the one who looks into the perfect law and follows its precepts is blessed. They not only hear but remember as well.

Proverbs 4:20–21 says, "Be attentive to my words; incline your ear to my sayings. Let them not escape from your sight; keep them within your heart." How can we spur our brains to remember well?

One of the best ways to etch God's Word into our memory is through music, which can also be restorative and healing. As I watched the adolescent with an epidural hematoma slowly recover, after several days, soft music began playing in the background. At first, I thought it was just to soothe visiting family. Later, I found out it was part of her music therapy.

Psalm 100:1–2 instructs us to "make a joyful noise to the Lord . . . ! Come into his presence with singing!" Singing brings joy, but hearing music hardwires connections within our brains. The prefrontal cortex not only is responsible for memory retrieval but also serves to process music. When we listen to our favorite praise songs, vivid memories remind us of where we were and even what we were doing when we first heard them.

> Create in me a clean heart, O God,
> and renew a right spirit within me.
> Create in me a clean heart, O God,
> and renew a right spirit within me.
> Cast me not away from thy presence, O Lord.
> And take not thy Holy Spirit from me.
> Restore unto me the joy of thy salvation,
> and renew a right spirit within me.

The words of Psalm 51:10–12 are engraved in my mind, but not because I ever took time to memorize them. During a youth retreat, God used a prayerful worship song to connect with my soul. Maybe you heard the notes in your head as you read the words. I can still feel the crackle of the campfire each time I hear the lyrics. The psalmist's words will forever be linked to the notes of the song for me.

Implicit memories are automatic and unconscious. They are formed when events, emotions, experiences, or even songs are linked to our memory. When we hear the melody of a meaningful worship song, we easily remember the words without looking at the lyrics. Alzheimer's patients can retrieve forgotten memories by listening to music they heard a time long ago. For many, the worship doxology is an implicit memory: the words flow from our mouths as soon as the first notes are played.

Not only is hearing music linked with memory but research has demonstrated various other positive effects associated with music. Listening to music reduces anxiety and symptoms of depression.[15] It disarms anger and antisocial behaviors while providing greater fulfillment in life.[16]

Paul encourages believers to speak to one another "in psalms and hymns and spiritual songs, singing and making melody to the Lord" (Eph. 5:19). The more we listen to uplifting songs of worship, the more implicitly we know that God is good, his mercies last forever. The lyrics of timeless hymns remind us of the promises of the Lord. Their poetic words speak of his infinite glory, honor, and power.

When we sing corporately, the sound of praise rises to the heavens. A multitude of heavenly hosts joins in when they hear our worship. The chorus of voices stirs our hearts and invigorates our minds. Martin Luther once described this phenomenon: "At home, in my own house, there is no warmth or vigor in me, but in the church when the multitude is gathered together, a fire is kindled in my heart, and it breaks its way through."[17]

Secular research tells us what we already know about

engaging in praise and worship music. Hearing the sweet melodies of gospel music decreases anxiety about death while increasing self-esteem and satisfaction in life.[18] These songs serve as a source of strength and provide meaning in the face of suffering.

Every Christmas, the majestic "Hallelujah Chorus" of Handel's *Messiah* engages people's auditory senses while strengthening their faith. When the promises of salvation reverberate through our eardrums together with a harmonious symphony of instruments, who could doubt the words of 1 Corinthians 15:55–57: "'O death, where is your victory? O death, where is your sting?' The sting of death is sin, and the power of sin is the law. But thanks be to God, who gives us the victory through our Lord Jesus Christ."

When we embrace sound with the Creator's intent, we begin to hear what God has in store for us. Through the reign of the Holy Spirit, every sound, voice, and noise is an opportunity to live as Christ did on earth. The Word made flesh teaches us that our ears are made for the Lord. As dwelling places for God, we reflect the image of God when we use the sense of hearing to help others worship the Lord.

It's true not only for hearing but also for each of our other senses: sight, smell, taste, and touch.

QUESTIONS FOR REFLECTION

1. Do you know someone who is hard of hearing? If so, what impact does hearing loss have on this person and their loved ones?
2. From the skin and cartilage of the outer ear to the cochlea, vestibule, and labyrinth of the inner ear, the auditory

system is a work of art. How can you protect your ears from hearing loss?

3. Destructive noises can be physical and spiritual. What external and internal voices do you struggle with the most?
4. John 8:47 says, "Whoever is of God hears the words of God." In what ways are you listening to the Word of God?
5. In Romans 2:13, Paul says, "For it is not the hearers of the law who are righteous before God, but the doers of the law who will be justified." How is the Holy Spirit convicting you to be a doer of the Word?
6. How can you speak to your neighbors with psalms, hymns, spiritual songs, singing, and melodies to the Lord (Eph. 5:19)?

NOTES

1. Christie, J. (1987). Helen Keller. In *Gallaudet encyclopedia of deaf people and deafness*. McGraw-Hill.
2. Gilsdorf, J. R. (2018). Into darkness and silence: What caused Helen Keller's deafblindness? *Clin Infect Dis* 67(9), 1445–49.
3. Love, J. K. (1933). *Helen Keller in Scotland: A personal record written by herself*. Methuen.
4. Gottlieb, M. I., & Williams, J. E. (1987). *Textbook of developmental pediatrics*. Plenum Medical Book Co.; Centers for Disease Control and Prevention. (n.d.). What is hearing loss in children? Retrieved August 23 2022, from https://www.cdc.gov/ncbddd/hearingloss/facts.html
5. Centers for Disease Control and Prevention. (n.d.). Hearing loss treatment and intervention services.

Retrieved August 23, 2022, from https://www.cdc.gov/ncbddd/hearingloss/treatment.html

6. Heffner, R. S., & Heffner, H. E. (1982). Hearing in the elephant (*Elephas maximus*): Absolute sensitivity, frequency discrimination, and sound localization. *J Comp Physiol Psychol* 96(6), 926–44.
7. Bailey, B. J., & Johnson, J. T. (2006). *Head and neck surgery—otolaryngology.* Lippincott Williams and Wilkins.
8. Welling, D. R. U. (2017). *Fundamentals of audiology for the speech-language pathologist.* Jones and Bartlett Learning.
9. Centers for Disease Control and Prevention. (n.d.). What noises cause hearing loss? Retrieved August 23, 2022, from https://www.cdc.gov/nceh/hearing_loss/what_noises_cause_hearing_loss.html; National Institutes of Health. (n.d.). Noise-induced hearing loss." Retrieved August 23, 2022, from https://www.nidcd.nih.gov/health/noise-induced-hearing-loss
10. Carroll, Y. I. et al. (2017). Vital signs: Noise-induced hearing loss among adults—United States 2011–2012. *MMWR Morb Mortal Wkly Rep* 66(5), 139–44.
11. Centers for Disease Control and Prevention. (n.d.). What noises cause hearing loss? Retrieved August 23, 2022, from https://www.cdc.gov/nceh/hearing_loss/what_noises_cause_hearing_loss.html.
12. Centers for Disease Control and Prevention. (2020, January 6). Too loud! For too long! Retrieved August 23, 2022, from https://www.cdc.gov/vitalsigns/hearingloss/index.html
13. McCrindle Research. (2022). Changing faith landscape in Australia.

14. Clark, S. et al. (2008). Car surfing: Case studies of a growing dangerous phenomenon. *Am Surg 74*(3): 260–61.
15. Evans, M. M., & Rubio, P. A. (1994). Music: A diversionary therapy. *Today's OR Nurse 16*(4), 17–22; Erkkila, J. et al. (2011). Individual music therapy for depression: Randomised controlled trial. *Br J Psychiatry 199*(2), 132–39; Chan, M. F. et al. (2012). Effects of music on depression in older people: A randomised controlled trial. *J Clin Nurs 21*(5–6), 776–83.
16. Baker, F., & Bor, W. (2008). Can music preference indicate mental health status in young people? *Australas Psychiatry 16*(4), 284–88.
17. Rayburn, R. G. (2010). *O come, let us worship*. Wipf and Stock.
18. Bradshaw, M. et al. (2015). Listening to religious music and mental health in later life. *Gerontologist 55*(6), 961–71.

PART 2

EMBODIED WORSHIP

The gift of sense allows us to perceive the world around us. Sight, hearing, taste, touch, and smell invite us to stand in awe of creation. Through the sense organs we experience the beauty of snowcapped mountains, the sound of praise, the taste of chocolate chip cookies, the smell of fresh coffee, and the caress of loved ones. Each one beckons us to glorify God in refreshing ways.

Similarly, we were created to breathe, move, create, rest, and love. These bodily functions facilitate engagement with the world while helping us navigate interpersonal relationships. It is natural to worship through traditional disciplines such as fasting, prayer, and meditation. Yet we can also glorify God through our hearts, lungs, minds, movements, and in our rest.

When we understand that God created the systemic organs as a way for our bodies to accomplish his purposes, we begin to see their functions in a different light. Instead of playing mundane roles within the body, they help us embrace God's vision for our lives while reflecting the image of God to others.

In the following chapters, we will examine how to embody worship through breathing, movement, creativity, rest, and love. An understanding of the innate biological processes that underlie each will encourage us to stay healthy. By realizing that spiritual and physical health are intimately related, we will be better positioned to treat our bodies as temples of the Holy Spirit.

6

BREATHE

Baby Logan, nestled comfortably inside his mother's womb, was vigorous, growing, thriving, despite his mother's diabetes. Regular checkups and medication kept his mother's blood glucose levels in a reasonable range during her otherwise uncomplicated pregnancy. The family prepared for Logan's arrival as any family would, purchasing a crib, car seat, and what seemed like a global supply of diapers.

But a few weeks short of Logan's due date, his mother's water broke. Logan's mother frantically packed her bags while his father prepared to drive her to the hospital. As the on-call pediatrician that evening, I was aware of this diabetic mother and the potential for complications. Babies born to diabetic mothers can exhibit abnormal insulin levels affecting the production of important survival elements such as surfactant. Even the stress of premature delivery could threaten that delicate balance of glucose and insulin in the mother's body.

Surfactant is a remarkable mixture of fats and proteins that lowers the surface tension between air sacs in the lungs. Lining the lung tissue, surfactant prevents the lungs' tiny air sacs—alveoli—from collapsing during exhalation. That protection makes it easier for newborns to breathe after delivery.

But surfactant isn't produced until about twenty-six weeks gestation. Sometimes, premature babies do not have enough lining their lungs at birth. In these cases, babies are prone to respiratory distress syndrome (RDS), which occurs in 1 percent of all pregnancies.

Though only a few weeks premature, Logan had increased risk factors for RDS. The nurses paged me to the delivery room after his mom had progressed through the initial stages of contractions for several hours. As I entered the birthing suite, a palpable sense of anxiety and anticipation filled the air. I stood near the radiant warmer awaiting Logan's delivery as the OB doctor worked with his mother.

"Push, push, push!"

As the OB doctor urged the mother, I monitored Logan's heart rate, which remained steady and robust throughout. Logan's mother was visibly tired but ready to give birth. She had been pushing for an extended period, and it was that time. She mustered all her strength with the next contraction, forcing Logan completely through the birth canal.

I listened for a vigorous cry, but all I heard was silence. Logan's body was limp, showing no signs of movement. Though his skin was covered in birth secretions, his extremities looked decidedly purple. Years of pediatric training reverberated through my mind as I reached for a bag valve mask and asked the assisting nurse to prepare the correct-sized endotracheal intubation blade.

After laying him on the radiant warmer, I positioned Logan's head in a neutral position with his neck slightly extended. Then I suctioned the back of his mouth and throat, careful not to overstimulate his vagus nerve, the most important nerve in the body. I was fully prepared to intubate him.

Miraculously, these maneuvers awakened Logan. He gave a shrill cry and began grunting. Several rescue breaths were administered through the bag valve mask as Logan's breathing and color slowly but steadily improved.

As a precautionary measure, he was transferred to the neonatal intensive care unit for further care. Though ventilation wasn't needed, Logan benefited from additional oxygen support. In time, he recovered fully from RDS and began breathing normally without assistance.

Breathing is life.

BREATH IS LIFE

Hebrew writers chose the word *neshama* to represent this invisible air we call breath. Though other words have the same connotation, only *neshama* is applied to both God and man. Humanity alone receives the divine breath of God, no other created beings.

With this gift, we move from lifeless bodies to Spirit-filled beings created in the image of God. Not only does *neshama* represent life-giving breath but it is a metaphor for our souls. It points to the centrality of breath in our lives. That same breath disappears with death.

At every healthy birth, newborns emerge from the womb with breath. Before that time, their lungs are devoid of air, filled with liquid. A fetus in utero is enclosed by an amniotic sac, a thin-walled membrane surrounding a baby throughout pregnancy. This jellylike pouch is full of amniotic fluid, which protects infants from injury, helps regulate their temperature, and allows immature lungs to form.

While every other bodily organ is functioning weeks before birth, the lungs lie dormant, waiting to support life. Mothers hear the heartbeats of their babies while sensing the movement of their limbs at both welcomed and inopportune times. Classical music engages the brain while helping improve memory retention, even during the early stages of pregnancy. Kidneys churn out urine, which is deposited in amniotic fluid, then swallowed by infants repeatedly. Only the lungs remain idle.

But in an instant, they awaken as newborns transition from a mother's womb to the outside world. Contractions move the infant through the birth canal while pushing amniotic fluid out of its lungs. As a baby leaves the mother's body, it gasps for air. Oxygen replaces amniotic fluid. With fierce determination, the lungs inflate for the first time while the remaining amniotic fluid is resorbed. Life-giving gas exchange occurs as oxygen moves into the bloodstream and carbon dioxide is eliminated from the blood.

An infant's precious first cry signals their triumphant entry into the world. Though piercing to first-time parents, vigorous cries indicate healthy lungs and adequate oxygenation. Dead silence is a harbinger of hypoxemia for pediatricians in attendance, requiring invasive maneuvers for resuscitation. In the first few minutes after birth, time is fleeting. Asphyxia—a deficit of oxygen—can severely affect newborns.

For healthy babies, breathing is spontaneous. Every breath is effortless, controlled by the autonomic nervous system. Like clockwork, infants inhale and exhale without cognition. Whether they are asleep or awake, their respirations occur freely. Unless they succumb to illness, each inhalation provides the perfect amount of oxygen to sustain their vital organs. With every exhalation, the body eliminates harmful gases.

Yet as these babies grow into adults, the natural, unobstructed breathing cycle isn't always a given.

THE DEATH OF BREATH

In 2023, the submersible *Titan* lost contact with its support vessel one hour and forty-five minutes after descending. On its way to tour the wreckage of the iconic cruise ship *Titanic*, the lost sub was more than halfway to the bottom of the ocean. The next day, the Coast Guard estimated that the submersible had between seventy and ninety hours of oxygen left. A frantic search-and-rescue operation ensued. It ended with a subsea remote-operated vehicle finding debris from *Titan*'s pressure hull on the seafloor at a depth of 12,500 feet. After conducting further research, the Coast Guard determined the *Titan* likely experienced a catastrophic implosion after departing on its voyage, so it's unlikely the victims experienced the torture of oxygen loss that many feared they did.[1]

The body can last for only a few minutes without oxygen. In the absence of air, we begin to experience headaches, increased heart rate, shortness of breath, confusion, and eventually loss of consciousness. Brain damage is irreversible.

Respiratory illnesses and environmental factors can restrict our ability to breathe. These diseases affect the airways that carry gases, the blood vessels within the lungs, and the structure of the organ tissue. Chronic diseases like asthma and emphysema cause suffering and debilitation. Pneumonia and infectious diseases like COVID-19 sometimes require hospitalization. Many are inherited genetically. Some are self-inflicted secondary to smoking or the inhalation of illicit substances.

Though the autonomic nervous system regulates breathing, input via voluntary control can derail natural rhythms. Both autonomic and cortically controlled breathing patterns respond to environmental stimuli. Illness, trauma, pathology, and psychosocial disorders can disrupt our respiratory drive. When this occurs, other bodily organs are at risk because of a scarcity of oxygen.

Though babies breathe naturally, adults can sabotage the rhythms of breathing or destroy the breath of life within others.

Such disruption results from specific actions or inaction, choices that directly affect others or indirectly impact them through the environment. In the most egregious cases, violence results in death, like Cain's killing of Abel. More recently, the death of George Floyd was marked by his dying words, "I can't breathe." His final plea for help underscores our frailty without one of our most precious commodities: the air we breathe.

INTENTIONAL BREATHING

In the face of fear, anxiety, or stress, we typically breathe too fast. Rapid breathing is a natural result of the fight-or-flight response, controlled by the sympathetic nervous system. The cascade of symptoms of stress-related breathing can lead to heart disease, changes in hormone levels, increased inflammatory markers, and even chronic fatigue over time. Extreme cases result in hyperventilation, rapid breathing that upsets the balance of the pulmonary system. When we exhale too fast, an excess of carbon dioxide is excreted from the body. This leads to the narrowing of blood vessels that supply oxygen to vital organs. With time, confusion and light-headedness set in.

That's what happens to Cindy every time she experiences panic attacks. It starts with innocent worry. Cindy's fear of being around people, anthropophobia, is heightened in crowded public places. Typically, psychological coping mechanisms help her face these fears. But with the rise of hate crimes, Cindy, an Asian American female, is more prone to panic attacks. When they occur, something as instinctive as breathing is suddenly difficult. Every breath is a battle against her body, with each respiration growing more and more restricted. Instead of the natural, rhythmic waves associated with spontaneous breathing, she hyperventilates with short, abrupt breaths while convinced she can't breathe.

Whether symptoms of hyperventilation are barely noticeable or life threatening, healthy breathing practices can help individuals cope. Conscious and purposeful breathing can positively affect reactions to stress and anxiety while mitigating chronic pain, immune and digestive disorders, asthma, and other illnesses. By training our lungs to breathe well, we redeem our breath for the Lord and return one step closer to what God intended.

NIH researcher Esther Sternberg describes the benefits of conscious breathing. "The relaxation response is controlled by another set of nerves—the main nerve being the vagus nerve. Think of a car throttling down the highway at 120 miles an hour. That's the stress response, and the vagus nerve is the brake," says Sternberg. "When you are stressed, you have your foot on the gas pedal to the floor. When you take slow, deep breaths, that is what is engaging the brake."[2]

Diaphragmatic breathing (also known as abdominal breathing or belly breathing) is a helpful technique to improve health. It facilitates relaxation, lowers blood pressure, and decreases stress levels. This deep, conscious breathing keeps the body and mind functioning optimally while also enhancing focus or concentration. It is the basis for a plethora of contemplative practices.

The diaphragm—a dome-shaped muscle located directly below the chest—helps facilitate respiration. Normal breathing causes the chest to expand, while diaphragmatic breathing results in bulging of the abdomen. Vocalists often utilize diaphragmatic breathing to enhance their singing. Stronger diaphragmatic muscles help them sustain their voices for extended periods. Opera singers can routinely hold notes for several minutes, much longer than a person untrained in abdominal breathing.

Whether we engage in diaphragmatic breathing or other purposeful techniques like thoracic breathing or alternate-nostril breathing,[3] healthy breathing allows us to redeem our bodies. Thoracic breathing focuses on expanding the rib cage and intercostal muscles during inhalation, with a return to normal during exhalation. Alternate-nostril breathing splits complete cycles of inhaling and exhaling between the right and left nostrils. By embracing the art of conscious breathing, we decrease heart rate, lower blood pressure, and improve exercise tolerance. Over time, lung function and cardiovascular function gradually improve.

INHERITING BREATH

Healthy breathing benefits our bodies while also calming our hearts and minds. We relax and slow down, inviting opportunities for intimacy with God. Christian engagement in the art of breathing—with deep and historical roots in Eastern religions—has been limited. Some evangelical leaders condemn practices associated with Eastern meditation. They argue that Christians cannot develop these disciplines without disregarding biblical truth and witness.[4] Douglas Groothuis, professor of philosophy at Denver Seminary, warns that "all forms of yoga involve occult assumptions."[5] Is it possible to uncouple healthy breathing practices from the occult? As a devoted Christian, physician, and pastor, I believe it is.

For thousands of years, sages, mystics, and wise men realized the primary physiologic benefits of proper breathing. Many used the power of breathing techniques to help control their minds, shift their consciousness, and provide healing for their bodies. They recognized that breathing encompasses more than primary respiratory cycles. There is something sacred tied to the secular.

The ancient Indian system of yoga has enjoyed widespread popularity in the twenty-first century. *Pranayama* is the Sanskrit word describing yoga breathing exercises or breathwork.[6] *Prana* means "breath, life force, and vital energy." *Pranayama* is thus the expansion of breath, life force, and energy. Yogis have been practicing and perfecting breathwork exercises for more than five thousand years. These include states of posture, movement, technique, and regulation of breathing.

Qigong, literally "energy work," is the cultivation of

life energy through breathing, movement, and meditation. The origin of qigong dates back to the legendary Chinese Yellow Emperor in 2696 BC, while scholars acknowledge Confucius and Mengzi's influence. Hundreds of systems of qigong attempt to harness willpower, focus, and energy. The coordination of breath with movement and meditation presumptively drives *qi* (life force) through intended pathways while opening up unintended blockages.

Though yoga and qigong share common Eastern backgrounds of Hinduism, Buddhism, and Jainism, the centrality of breathing and airflow reverberates throughout Christendom as well. The Hebrew word *neshama* is typically described as the breath of life or act of breathing. *Ruah* most often means wind or movement of air. *Nefesh* has the connotation of rest, defined as moving air coming to rest upon humanity.

As early as Genesis 1:2, we witness the Spirit of God hovering over the face of the waters. The Spirit of God then breathed life into the nostrils of man, and he became a living being. Breathing was the very act of inheriting God's breath within our bodies. We received not only this physical breath but also our spiritual souls.

In the New Testament in Acts 2, the Greek word *pneuma* describes the Holy Spirit at Pentecost. Yet it also represents wind in John 3:8 and breath in Revelation 11:11. As Christians, we receive the breath of life from the Father (Acts 17:25) and the gift of the Spirit (John 7:39). Instead of being drunk with wine, Paul encourages us in Ephesians 5:18 to be filled with the Holy Spirit. This gift of the Spirit is associated with Jesus' breath. Following his resurrection, Jesus breathes on the disciples while inviting them to receive the Holy Spirit (John 20:22).

New Testament scholar Luke Timothy Johnson contends that Christianity has skewed too far toward emphasizing external religious expressions.[7] This has left a void within the internal substance of faith—the heart's inner experience and devotion. Internal dimensions of faith focus on ascetic practices like prayer, fasting, and meditation. For centuries, these internal elements of faith were expressed most fully by mystics.

As Christians, we are wise to establish boundaries between our faith and the practices of other religions. Yet we are foolish not to fully embrace the benefits of internal spiritual disciplines, which allow us to flourish in our relationship with God in daily life. It is easy to fall into syncretism, blending worldviews without caution. Therefore, we must be mindful when we engage in stereotypic Eastern religious practices. Yet I believe it is possible to separate the health-related aspects of yoga or qigong from their postures designed to focus on the divine.

We do this by marrying the physiologically proven benefits of breathing exercises to Christ-centered spiritual practices such as prayer, meditation, fasting, and worship. Following are examples of breathing practices rooted in Christian tradition that offer physical and spiritual benefits.

BREATH PRAYER

By integrating healthy breathing practices with timeless spiritual disciplines, we embody worship. Spiritual disciplines such as meditation are invitations to dwell on the Word of God while casting out the distractions of the world. Connecting our minds and bodies allows a heightened sense of awareness and singular devotion.

With roots dating back to Eastern Orthodox Christians, breath prayer is an ancient spiritual discipline that combines healthy breathing with thoughtful prayer. A form of contemplative prayer, it focuses our minds on God. In tying short prayers to inhaling and exhaling, Christians can pray without ceasing (1 Thess. 5:17).

The art of breath prayer involves drawing deep, full breaths through diaphragmatic breathing. These slow, deliberate breaths have a plethora of scientific benefits while preparing our hearts and minds for prayer. Inhaling is often focused on the Godhead (Father, Son, and Holy Spirit), while exhaling reflects a spiritual desire or truth. A large variety of meaningful words and phrases resonate with different people:

- "[Inhale] Be still and know . . . [exhale] that I am God."
- "[Inhale] Sophia . . . [exhale] wisdom."
- "[Inhale] Jesus . . . [exhale] companion."
- "[Inhale] Abba Father . . . [exhale] I love you."
- "[Inhale] Jesus Christ . . . [exhale] show me your truth."

Breath prayer takes away natural tendencies to obsess on the meaning of words and delivery of language within our prayers. Prayer becomes more natural, as constant as breathing. The spontaneous aspect of breathing turns our attention to the presence of God.

One of the earliest breath prayers emanates from the story of the blind beggar in Mark 10:47. As Bartimaeus calls upon the Lord to heal him, he says, "Jesus Christ, Son of David, have mercy on me!"[8] In this case, inhalation is tied to the incarnation, while exhalation beseeches mercy. Repetition of these words allows constant attention to God's presence, which provides the peace of Christ.

Hebrew and Greek words represent both the Spirit of God and the air we breathe. Breath prayer allows Christians to receive the Spirit of God while uttering simple requests before him. Symbolically, we visualize the breath of life from Genesis 2:7 and the gift of the Holy Spirit from John 7:39 while we surrender our lives to him. By embodying our prayers, we realize the truth of Acts 17:28, that in him we live, and move, and have our being.

SPIRITUAL BREATHING

Similar to breath prayer, spiritual breathing can foster intimacy with God. Bill Bright, the founder of Cru, describes it as "a process of exhaling the impure and inhaling the pure, an exercise in faith that enables you to experience God's love and forgiveness and walk in the Spirit as a way of life."[9]

The discipline of spiritual breathing involves the same physical breathing components as breath prayer: inhaling and exhaling. As Christians, we inherit the breath of life, but also the Spirit of God. While being filled with the latter, we implore God to control our lives and protect us from sin. As we inhale spiritually, we receive the fullness of God in our lives. It is a repetitive but meaningful process whereby we continuously ask the Holy Spirit to empower us with every breath we take.

In contrast, the process of spiritual exhalation is one of confession of sin while claiming forgiveness in Christ. James 5:16 urges us to confess our sins to another. With every exhalation, we acknowledge our imperfections and fallenness while repenting of our sins. When we embrace the habit of spiritual breathing, every breath we take helps us to focus more on the kingdom of God and less on our circumstances.

BREATH AND VOICE

In 1986, eighteen-year-old Ronda Morrison was brutally murdered in Monroeville, Alabama. A horrified town stood in shock as local police were unable to solve the crime. Though he had no criminal record, forty-five-year-old Walter McMillian was tried for capital murder. At the time of Ronda's murder, witnesses placed Walter eleven miles away from the crime scene with his family. Nevertheless, he was convicted and eventually sentenced to death by electrocution.

Two years later, Bryan Stevenson took on the case post-conviction and began working to appeal the verdict and sentence. His team of lawyers found ample evidence for McMillian's innocence, including tape recordings proving that the state's lone eyewitness had been coerced to lie under oath during the trial. In 1993, Alabama's Court of Criminal Appeals overturned Walter's conviction. He was set free after serving six years on death row.

Stevenson eventually created the Equal Justice Initiative, a human rights organization dedicated to helping the poor, the incarcerated, and the condemned. He and his staff won the release or reversal of conviction for more than 135 wrongly convicted prisoners on death row and relief for hundreds of others. Stevenson and his fellow lawyers spoke for those who could not speak for themselves. By discerning their cases, they defended the rights of the prisoners and needy among us. They prevented the voices of these people from being swallowed by death in silence.

As temples of the Holy Spirit, we embody worship by connecting spiritual disciplines to the natural process of breathing. But how do we sacrifice our breath for the sake of others? In speaking for those without a voice, we give breath to the vulnerable. By ensuring their voices are heard, we guarantee their cries don't remain unnoticed.

The author of Proverbs urges us to speak for those who cannot speak for themselves, for the rights of those who are destitute (31:8). Who are those people without a voice? They are the poor, the homeless, the oppressed, the foreigner, and the widow (Ex. 23:9; Lev. 19:33; Ps. 68:5; Isa. 1:17; James 1:27). Time and time again, God shows compassion to the least of these. Moses cautions the Israelites not to take advantage of widows and orphans or to oppress sojourners. Leviticus teaches them to leave the fallen fruits of their vineyards to the alien and poor. Every seventh year, Israelites allowed their land to remain unplowed so the destitute could harvest food and eat from it.

How do we give voice to the vulnerable? According to Proverbs 31:9, we speak up and judge fairly, defending the rights of the poor and needy. Globally, inequalities and inequities abound. They emanate from psychosocial determinants that result in disparate environmental conditions and societal opportunities. Such disparity leads to unequal distribution of resources, wealth, and even punishment.

The process of judgment involves discernment in addition to decision-making. By discerning the needs of the vulnerable among us, we realize how to speak for them. Our voices defend their rights and begin to rectify their oppression. In giving our breath to the speechless, we embrace the clarion call of Micah 6:8 to act justly and love mercy, to walk humbly with our God.

In Psalm 150:6, David reminds us that everything that has breath should praise the Lord. We praise him in his sanctuary and within the firmament of his heavens. The heavens declare the glory of God; the skies proclaim the work of his hands (Ps. 19:1). With every breath we take, we worship the Father. We pray with our voices. We exalt from our lungs. Everything that has breath should praise the Lord, but every breath we take can be a sacrifice for others. By speaking for those without a voice, we regard them as temples of the Holy Spirit. In the discernment of inequity, we defend their unspoken needs. In doing so, we move one step closer to the vaults of heaven and the throne of the Lamb.

QUESTIONS FOR REFLECTION

1. Humanity alone receives the breath of life from God, which transforms us from lifeless bodies to Spirit-filled beings. How have you taken for granted the breath of life?
2. We are reminded of the frailty of life each time we take a breath. In what ways can you sabotage your breathing or that of others?
3. Intentional breathing can positively affect reactions to stress and anxiety while mitigating chronic pain, immune disorders, asthma, and other illnesses. Take a moment to practice diaphragmatic breathing.
4. Breath prayer allows us to pray without ceasing (1 Thess. 5:17). When inhaling, we focus on the Godhead. While

exhaling, we consider a spiritual truth or desire. Spend the next few minutes in breath prayer.

5. Psalm 150:6 reminds us that everything that has breath should praise the Lord. How is the Holy Spirit convicting you to glorify him with your lungs?
6. In what ways might you speak for those without a voice?

NOTES

1. Sanchez, R. (2023, June 4). The unsettling days after the *Titanic* submersible's demise. CNN. https://www.cnn.com/2023/06/24/us/missing-titanic-submersible-timeline/index.html
2. Cuda, G. (2010, December 6). Just breathe: Body has a built-in stress reliever. National Public Radio. https://www.npr.org/2010/12/06/131734718/just-breathe-body-has-a-built-in-stress-reliever.
3. Ma, X. et al. (2017). The effect of diaphragmatic breathing on attention, negative affect and stress in healthy adults. *Frontiers in Psychology 8*, 1–12.
4. Mohler, A. (2010, September 20). The subtle body: Should Christians practice yoga? https://albertmohler.com/2010/09/20/the-subtle-body-should-christians-practice-yoga
5. Groothuis, D. (1986). *Unmasking the new age*. InterVarsity Press.
6. Roach, M., & McNally, C. (2005). *How yoga works*. Diamond Cutter Press; Satchidananda, S. (2012). *The yoga sutras of Patanjali*. Integral Yoga Publications.
7. Johnson, L. T. (2010, February 22). Dry bones: Why religion can't live without mysticism. *Commonweal*. https://www.commonwealmagazine.org/dry-bones

8. Though in Scripture Bartimaeus says, “Jesus Christ, Son of David,” Christians pray “Son of God” instead.
9. Bright, B. (1999). How you can walk in the Spirit. CRU. https://www.cru.org/us/en/train-and-grow/transferable-concepts/walk-in-the-spirit.html

7

MOVE

For many, running a marathon is among the greatest achievements of a lifetime. Untrained adults can run hardly more than a mile or two before taking a break. Nonrunners have a difficult time finishing two miles in less than twenty minutes. But even for active road racers, completing 26.2 miles can fatigue the heart, lungs, muscles, and joints to the point of injury.

Marathon runners have collapsed because of dehydration, hyperthermia, exhaustion, and emotional stress. Near the finish line, lactic acid builds up in the muscles, leading to abnormal heart rhythms and exacerbating underlying heart problems.

Yet on October 12, 2019, Eliud Kipchoge ran at a clip of 13.1 mph for 26.2 miles on a track in Europe, becoming the first person to run a sub-two-hour marathon. His final time—one hour, fifty-nine minutes, forty seconds—was equivalent to running each mile in four minutes, thirty-five seconds![1]

For Kipchoge to sustain such mind-boggling speeds, he needed to maintain an oxygen uptake, or VO_2, of approximately four liters per minute. Breathing in this amount of oxygen for two hours is impossible for ordinary people. Kipchoge metabolized oxygen twice as fast as the average runner while maintaining sprint speed.[2]

His ability to run at high speeds without fatigue for so long speaks to the wonder of God's creation, where physiology, genetic makeup, and perseverance result in miraculous feats. Though most of us will never run a sub-two-hour marathon, we can still optimize our running economy and lactic acid threshold. We can do more than we imagine. With dedication and training, our physiology is highly adaptable.

By training the body, a relatively healthy person can soon run for thirty minutes without a break. Larger amounts of oxygen will be consumed while our bodies metabolize it to generate energy. Slowly, aerobic exercises like running, swimming, and biking become second nature. Walking up a flight of stairs no longer leaves us breathless.

The more we move, the stronger our hearts and lungs become. By burning more calories, we keep weight gain at bay. Exercise gets rid of fat while creating fuel for our bodies. Regular activity lifts our mood, improves concentration, and leads to better sleep. Physical exertion even lowers the incidence of diseases like colon and breast cancer.[3] It is a prescription worth its weight in gold.

With the psalmist, we should raise our voices to God: "I praise you, for I am fearfully and wonderfully made" (Ps. 139:14).

DESIGNED TO MOVE

The human body was created to move. Even before developing purposeful movements, infants possess involuntary reflexes. A baby's lack of movement is a red flag for parents and physicians.

The Moro reflex, or startle reflex, is part of every newborn's first physical exam.[4] This primitive reaction is elicited by raising the head of a lying infant approximately thirty degrees and then suddenly dropping the baby while supporting its body to prevent injury. The baby will immediately startle, moving back their head while extending both arms and legs. Within a split second, they quickly draw their limbs back inward.

If you gently stroke the palm of an infant's hand, they will close their fingers in an apparent grasp. When you touch the corner of their mouth, the baby will turn toward that side while opening their mouth as if ready to eat. Even if they are many months away from walking, newborns appear to be taking steps when held upright with their feet touching the ground.

As children grow older, they develop gross and fine motor skills. The former allows them to engage larger muscles from the upper and lower body in coordination. Eventually, the synchronization of muscles, tendons, bones, and cartilage propels children to walk, run, bike, and swim. Fine motor skills of the hands and fingers enable the precise control of smaller objects needed for eating, drinking, writing, and a variety of activities for daily living.

From birth to adolescence, children refine their movements. Within the first two months of life, infants track parents' faces by turning their heads toward the sound of their voices. At around five months, they roll over for the first time. After crawling or scooting around for more months, they finally begin cruising while holding on to furniture. By the time they are fifteen months old or sooner, most babies are walking, with a keen eye toward the stairs.[5]

In the blink of an eye, toddlers can jump with both feet off the ground. By two or three years old, they're running and riding a tricycle. As little as a year later, they begin to skip, jump, and play hopscotch while learning to catch a ball. When kindergarten arrives, most are expert climbers. Some can already swim and ski better than their parents.

It's easy to take for granted the exquisite biomechanical engineering of our musculoskeletal system. Before suffering a grade 3 sprain to my ankle, I never considered how essential ankles are to everyday mobility. Unable to ambulate without crutches and a cumbersome boot, I found even mundane activities to be impossible. Pressing on the gas pedal was both painful and difficult, making it dangerous for others on the road or in the car with me, and severely limiting my ability to work and travel.

Ankle ligaments are like bungee cords that connect the feet and legs. Together, they provide stability for the ankle joints, specifically the subtalar joint, talocrural joint, and inferior tibiofibular joint. With exercise or in the aged, these ligaments can easily tear. Playing sports that require moving back and forth or rolling the foot often leads to ankle injuries. Inadvertent twisting of the ankle can result in mild strains and more severe sprains.

When functioning properly, the ankles provide extraordinary mobility. They stabilize our feet, allowing us to withstand one and a half times our body weight when walking and up to eight times our weight while running. Strong ankles help us stop on a dime while running. They provide flexibility, strength, and power for cutting, lunging, and jumping. Healthy ankles allow the body to move freely with effortless gait.

When we look at the intricacy of our bodies, we realize they are truly works of art. First Corinthians 12:12 alludes to the exquisite way the Creator has shaped every part of our bodies to function in unity, "for just as the body is one and has many members, and all the members of the body, though many, are one body, so it is with Christ."

The intricacy and connectedness of our bodies is awe inspiring. Every system works in unison but also in concert with the other systems' parts. God designed each of our bodily members to function in a complementary fashion. Ligaments connect the largest bones and help stabilize joints. Tendons attach bones to strong muscles that propel our bodies into motion. Together, they form a perfect symphony of motion that allows us to live, grow, thrive, and move.

Only an all-powerful creator could design our inward parts and perfectly knit us together in our mothers' wombs (Ps. 139:13). David writes that we are fearfully and wonderfully made (v. 14). The original language of the psalm affirms that each individual is made with great care—unique and set apart for God's purposes.

Every action in life, whether mundane or noteworthy, requires movement. So when our movement becomes limited for some reason, the ramifications can be tremendous.

FRACTURED MOVEMENT

As I walked into the modest one-story building in South Asia, it was easy to tell that this orphanage was far from ordinary. Crutches hung on all the walls. Wheelchairs lay in obvious need of repair. Unlike in most homes, the children were hardly moving. Many lay sprawled across their beds. Their limbs seemed frozen to their tiny bodies.

Some of these youngsters had never experienced what we take for granted: simply walking. They could only observe from a distance the joy of running around during recess. For many, their disability was born from inherited movement disorders or diseases like cerebral palsy. The rest of the kids with disabilities had succumbed to illnesses, injuries, and vaccine-preventable diseases like polio.

As I smiled and waved at the children, one little girl walked up to me using her bare hands. Though her leg muscles had atrophied long ago, she compensated by using her arms to ambulate. As she slowly approached me, I reached for a magic flower I had stashed away in my pocket. Instantly, the little flower blossomed into a bouquet of roses. A look of surprise and wonder came over her face when she saw the colorful roses.

The magic trick left the girl and her friends speechless, so I began telling Bible stories while integrating gospel magic and juggling. While sharing about the armor of God, I made animal balloons for each child. Most had never witnessed balloons twisted into such unique shapes. They giggled and laughed with utmost joy.

The orphans might have assumed that this was the extent of our hospitality. But in the ensuing weeks, we replaced their dilapidated wheelchairs with newer ones and physical therapists worked with their broken bodies. Whether they were young or old, newly infirm or disabled from birth, each of these children was robbed of the use of their muscles, bones, joints, and ligaments. In resource-poor countries, the chance to rehabilitate and regain lost function is slim. But for the first time, some of these children experienced hope through the love of Christ.

Globally, nearly one billion people live with disabilities in any given year. More than seventy-five million require daily use of a wheelchair.[6] Each suffers from physical and mobility impairments. Disabilities can affect the upper and lower limbs, sabotaging manual dexterity and ambulation and requiring children and adults to depend on artificial devices, prostheses, crutches, and wheelchairs to function normally.

When our bodies stop moving, death is hastened. At all times, the body is busy at work. The heart pumps blood to vital organs. The lungs oxygenate red blood cells. The kidneys process waste, while the stomach and intestines extract energy from foods. If one vital organ shuts down, the rest of the body cannot function properly.

Children born with cerebral palsy have significant difficulty ambulating while maintaining posture and balance. *Cerebral* concerns the brain, while *palsy* is a weakness or limitation of the muscles. It is the most common motor disability in early childhood. Severe forms relegate individuals to lifelong care and the need for special prostheses to walk.

For those afflicted with cerebral palsy, movement—stolen from them—is also needed to rehabilitate and strengthen them. Without physical and occupational therapy, muscles become rigid and immobile. Supportive treatments and interventions relax muscles, alleviate pain, and recondition the body. Early diagnosis, timely intervention, and sustained care can have a profound impact on quality of life.

Many disabled orphans in resource-poor settings never have the opportunity to receive life-changing services and interventions. As a result, they can hardly move their arms and legs. Over time, weakened muscles atrophy from disuse. A lack of resources halts these children's movement and cripples their spirits. Some are relegated to stationary beds for as long as they live.

But the opportunity to move can invigorate them and offer hope.

Aimee Mullins was born with two legs and feet, but she had both legs amputated at twelve months secondary to fibular hemimelia—a condition where the fibula is short or missing. Orthopedic surgeons told her parents she would never walk. But Aimee didn't listen to the prognosis. Instead, she refused to believe she would be handicapped for the rest of her life and defied the odds.

With the help of artificial prosthetics, her life became an inspiration for all with mobility impairments. She was a star track athlete at Georgetown and the first amputee to compete on a college track team. During the 1996 Paralympic Games, she set world records in the long jump and the 100 meter and 200 meter races.

Aimee refused to allow the loss of both lower limbs to define her. Not only did she become an athletic star, but she also blossomed into a successful model. One of the most iconic American portraits is of Aimee perched on top of hand-carved wooden legs with six-inch heels. In 1999, *People* magazine named her one of the fifty most beautiful women.[7]

On inspiring others, she once said in an interview with ABC, "I want a child who thinks, 'Wow, what can I do with my new leg?' It's factual to say I am a bilateral-below-the-knee amputee; I think it's a subjective opinion as to whether or not I am disabled because of that."[8]

Despite her body's perceived limitations, Aimee Mullins

longed for movement. She didn't allow her double leg amputations to steal her mobility. No matter how fractured our physical ability is, we too can embrace this natural inclination of our bodies to move.

REDEEMING OUR BODIES THROUGH MOVEMENT

In Acts 17:28, Paul quotes the Athenian poet Epimenides of Crete to show that our lives exist because of God alone, not because of the pantheon of deities that the Greeks worshiped.[9] He writes, "In him we live and move and have our being." Every individual lives, moves, and receives their being from God. Humanity exists because the Spirit of God breathed life into dry bones.

At the core of this statement, we realize that to live is to move, to move is to have our being, to have our being is to live. Each is closely tied with the others. All are gifts from above.

To live and have our being, we must move. From the earliest heartbeats in the womb to the last breaths taken on our deathbeds, we derive meaning from activity. The choices we make with our hands and feet, voices and speech, countenances and facial expressions have a direct impact on our lives and those of others.

By witnessing the movement of God, we learn how to redeem our lives. At the dawn of creation, the Spirit of God was moving over the waters (Gen. 1:2). The Hebrew root word *rhf* means to hover, shake, or tremble. While the earth was still formless, the Spirit was stirring, preparing for the handiwork of God.

According to Genesis, the Spirit of God is the first to move. Every movement is purposeful and thoughtful. Before Jesus returned to the Father, he promised to send the Counselor to guide humanity. That day occurred at Pentecost (Acts 2). Cloven tongues of fire fell upon the disciples while a mighty rushing wind blew with a thunderous sound. For the first time, Spirit-filled Christians began to identify how the Holy Spirit moves.

First Corinthians 6:19–20 says our bodies are a habitation for the Holy Spirit. Not only does the Spirit dwell in us, but he moves within us. It is the Spirit who convicts us to the point of repentance. As we acknowledge our sin, our hearts are purified. In times of discouragement, the Spirit comforts and assures our souls. The Holy Spirit is constantly speaking, guiding, and teaching believers. He sanctifies us as we live, move, and have our being in him.

The movement of the Holy Spirit points people toward Jesus. Christ came to earth so we might witness his redemptive sacrifice. His every action, word, and deed was meaningful. From his walking with the disciples to his penultimate sacrifice at Calvary, Christ teaches us to redeem our bodies for the sake of the kingdom. His life, death, and resurrection fulfill the will of the Father, demonstrating a life of perfect obedience.

While Jesus submits to the will of the Father, the hand of God orchestrates the lives of his creation. He commands the mountains and hills to sing in praise and the trees of the field to clap their hands (Isa. 55:12). Though we plan our ways, it is God who establishes our steps (Prov. 16:9). When we acknowledge him, he makes straight our paths (Prov. 3:6).

As Christians, we live, move, and have our being to

honor and glorify God. Since we were bought with a price, we are called to glorify God with our bodies (1 Cor. 6:20). Whatever we do, whether eating or drinking, we do for the glory of God (1 Cor. 10:31). Revelation 4:11 reminds us, "Worthy are you, our Lord and God, to receive glory and honor and power."

We honor and glorify God by stewarding our bodies for his purposes. When we cultivate a lifestyle of discipline, our bodies reflect his glory. When we embrace our bodies as temples of the Holy Spirit, we treat them as houses of worship. In time, every step we take and every move we make becomes kingdom minded.

One of the best ways to combat physical decline is planned, purposeful movement. Such movement redeems our bodies, bringing them one step closer to the garden of Eden, where the effects of disease, illness, and aging were unknown to humanity.

It's not surprising that movement and exercise offer numerous physical benefits. No medication or pill provides the building blocks for health and wellness that movement does. The list of health conditions and diseases that exercise prevents includes diabetes, arthritis, high blood pressure, heart disease, stroke, and many types of cancer.

Most people think marathon runners were created for long distances, that they were born lean and fast. While this may be true for some, it is really their consistent physical exertion that allows them to burn fat. The body uses both carbohydrates and fat as fuel sources. To convert fat to energy, a tremendous amount of oxygen is needed. Sustained aerobic training allows the cardiovascular system to deliver oxygen and metabolize fat. The enhanced oxygen uptake and delivery result in increased energy and endurance. As the cardiovascular system works more efficiently, more oxygen and vital nutrients are delivered to organs and tissues throughout the body.

Regular, high-intensity aerobic exercise even slows the aging process. Research shows that this type of sustained and intense movement results in appreciably longer telomeres—the end caps of our chromosomes—compared with people with inactive or even moderately active lifestyles.[10] As cells divide over time, telomeres become shorter and shorter. Individuals with shorter telomeres possess biologically older cells. When the telomeres decrease to a certain size, cells are no longer able to replicate. People with the highest levels of physical activity have telomeres nine years longer than those with sedentary lifestyles.

Keeping fit is a natural component of physical health, but it positively affects spiritual and emotional well-being as well. Regular exercise combats depression while improving mood. After a stressful day of work, physical activity stimulates the release of dopamine, serotonin, norepinephrine, and other brain endorphins, leaving individuals cheerful, relaxed, and less anxious.

Similar to its effect on the heart and lungs, exercise increases blood flow to the head, triggering the growth and repair of brain cells via brain-derived neurotrophic factor (BDNF).[11] Exercise is one of the best ways to halt the progression of Alzheimer's disease.[12] Even those not suffering from memory disorders report improved memory and sharpened thinking.

It is not unreasonable to recognize that by improving our mood, we are better equipped to rejoice in the Lord. A cheerful countenance leads to increased kindness and gentleness. When we are relaxed, it is easier to find peace while being more patient toward others. Improved concentration, better memory, and enhanced learning allow us to remember and meditate on the Word of God. In time, we become like the tree planted by streams of living water in Psalm 1, which yields fruit in its season.

RUN THE RACE

In our culture, only 23 percent of Americans aged eighteen and older meet the recommendations for physical activity, which include 150 minutes of moderate-intensity aerobic activity or 75 minutes of vigorous activity weekly, plus muscle-strengthening activities twice a week.[13]

With a plethora of physical, mental, and spiritual concerns affecting us, and an even greater plethora of benefits to exercise and movement, why are we still so averse to moving? Our natural inclination is toward idleness. But today, technological advances have provided us with even fewer reasons to move.

Compared with previous generations, we spend more time in environments requiring prolonged sitting. The computer age brought a shift from blue-collar to white-collar jobs. Artificial intelligence and robots have automated many tasks requiring hard labor. According to the American Heart Association, sedentary jobs have increased by 83 percent since 1950.[14]

A typical person in the United States spends half of their day sitting at a desk. Normal office workers are glued to their chairs for an average of fifteen hours.[15] This doesn't count the long commutes many people suffer each day. Not only have computers and smartphones contributed to inactivity, but the advent of electric bikes and electric scooters means that even these natural opportunities for movement are diminished.

Yet as we've seen, God designed us to move. Idleness leads to the decay of the house, according to Ecclesiastes 10:18. And just as we can become sedentary in our physical activity, so we can become complacent in our spiritual lives.

For Paul, a race is the perfect metaphor for the life of faith. In his second letter to Timothy, he writes, "I have fought the good fight, I have finished the race, I have kept the faith" (2 Tim. 4:7).

When Paul wrote to the Corinthian Christians, he knew they were familiar with the ancient Olympic Games. The games occurred every four years in Greece for more than a thousand years. People near and far came to witness the elite competition.

Watching Olympic athletes is always inspirational. Contestants give every ounce of effort as they compete against one other. Witnessing the games, you can sense their determination and passion. It is breathtaking to see their perfectly sculpted bodies bend, flex, stretch, and leap to outperform other elite athletes.

World-class athletes inspire us to cheer and applaud their efforts. When athletes are in the throes of defeat, we empathize all the more. Sometimes their extraordinary feats motivate us to join their chosen sport on an amateur level. We may never move so elegantly, but we can still experience the joy of running, biking, or skiing.

Perhaps that's why Paul compares faith to athletics. Every day can bring sorrow or joy. According to the psalmist, "Weeping may tarry for the night, but joy comes with the morning" (Ps. 30:5). How we respond has eternal significance.

The way we run the race determines whether our faith is genuine. That's why Paul says, "I do not run aimlessly; I do not box as one beating the air. But I discipline my body and keep it under control, lest after preaching to others I myself should be disqualified" (1 Cor. 9:26–27). Our exertion in running the race of faith in no way invalidates the grace of God. Rather, it confirms the grace that we freely receive, for faith without works is dead (James 2:26). God gives us the ability and asks us to exercise it.

When we run our spiritual races, we do so with purpose. Olympic athletes train with an eye on a gold, silver, or bronze medal. They do not run aimlessly. Instead, they race to win the prize, with great discipline and self-control.

As Christians, that's how we should run—wholeheartedly without regret. Yet we run not to win a medal but to obey God. While we endure the obstacles of the race, we continue to fix our eyes on Jesus, for he is the author and perfecter of our faith (Heb. 12:2). Though pain and suffering await, we press on with grit and endurance.

It's easy to get sidetracked when we set our eyes on anything other than Jesus. That's why our singular focus should be on Christ. We run because the Holy Spirit lives within us. It is this living, breathing Spirit who convicts our minds, stirs our hearts, and moves our bodies. Walking, running, swimming, biking, we fulfill the will of God.

Similar to Usain Bolt, we can be confident of victory. Widely considered the best sprinter of all time, Bolt won the 100 and 200 meter races an unprecedented three straight times in the Olympic Games. His eyes were always fixed on the finish line. Anyone could tell by the way he moved that in his mind, his races had already been won.

Unlike Bolt, we don't have to be confident in our

abilities. Instead, we can be sure that our race will never be lost because of the sacrifice of another. Christ endured the cross for all humanity. By his sacrifice, he has already won the victory for us in the past, present, and future.

The discipline of physical activity strengthens our faith. Exercise requires motivation to begin and dedication to finish. Along the way, we develop perseverance and mental toughness. Each step refines our disposition in affirming that "suffering produces endurance, and endurance produces character, and character produces hope" (Rom. 5:3–4). For believers, that hope does not disappoint "because God's love has been poured into our hearts through the Holy Spirit" (Rom. 5:5).

GO AND MAKE DISCIPLES

Movement is essential to life, and it is also crucial to fulfilling our calling. By embracing the mission of God, we participate in reconciling sinful humanity to God. At the heart of this mission is evangelism, which comes from the Greek *euangelion*.

The prefix of the word means "good." When attending a funeral, we typically hear a *eu*logy, or good word, about the deceased. The ending of the word means "angel" in English—the messengers of God. Evangelism is thus sharing the "good message" or "good news" of Jesus Christ.

In a sense, then, we can understand evangelism as "angel movement." According to Psalm 91:11, God will command his angels to guard us in all our ways. Angels praise and worship God (Ps. 148:1–2; Isa. 6:3; Heb. 1:6). They appear before him and serve him (Ps. 103:20; Job 1:6). They battle against the forces of evil (Dan. 10:13) and carry out God's judgment (Rev. 7:1) and sometimes deliver answers to our prayers (Acts 12:5–10). Angels are active and dynamic, always moving.

Likewise, our participation in evangelism must be active. By carrying out the Great Commission, we affirm our love for Jesus. There are five scriptural passages regarding the Great Commission: Matthew 28:18–20; Mark 16:15–16; Luke 24:46–48; John 20:21–22; and Acts 1:7–8. By far the most famous is the Matthew passage, where Jesus says, "All authority in heaven and on earth has been given to me. Go therefore and make disciples of all nations, baptizing them in the name of the Father and of the Son and of the Holy Spirit, teaching them to observe all that I have commanded you. And behold, I am with you always, to the end of the age."

At the core of this iconic passage is the Greek imperative to make disciples of all nations. The multiplication of disciples is accomplished by going, baptizing, and teaching. Each can be accomplished only with commitment, dedication, and action.

To make disciples, we must go! The verb emphasizes the participatory, physical activity of evangelism. Disciples don't grow on the branches of faith, they are grafted onto the tree of life through evangelism. Our role is to share the good news of the gospel. We defend our faith when we are asked difficult questions, but we convey good news not only by our lips but by our lives.

Perhaps the idea of communicating that good news to others seems daunting. The word *go* is a clue for us. To do this, we must participate in, train for, and even practice the art of telling the good news. Sometimes words don't come easily regarding matters of faith or our experiences with Christ. Our lives are difficult to share. But stagnation leads to paralysis and ineffective communication. If our efforts are periodic and haphazard, they lose their purpose as we struggle to be relevant to people the Holy Spirit places before us. And if our lives are misaligned with the message, we create an even larger problem.

Jesus' command is clear. The question is not merely whether to go but also to whom you have been called. There are more than 7,417 unreached people groups in the world, which means that 42.5 percent of the world's population is considered unreached.[16] Whether we are called to move to a foreign land or visit a neighbor across the street, the Great Commission underscores the "personal" aspect of the gospel.

Baptism in the name of the Father, Son, and Holy Spirit is the next step in making disciples. Through the incarnation of Christ, God paved the way for humanity to receive salvation. After Jesus died and rose again, he sent the Holy Spirit to fill the hearts of believers. It is this incarnation that allows us to live, breathe, and move for the kingdom.

The early church viewed baptism as participation in Christ's passion, death, and resurrection. Through immersion in water, believers plunge into the crucifixion of Christ and join in his death. But just as water is linked to death in the Psalms (42:7; 69:2), it is also tied to victory. First Peter 3:20–21 connects the salvation of Noah through the flood with Jesus' triumph over death communicated through baptism. Paul ties the crossing of the Red Sea to baptism in 1 Corinthians 10:1–2.

The sacrament of baptism requires engagement, both physically and spiritually. Baptism is the public adoption into faith. Believers profess their faith in a crucified, buried, and risen Christ while acknowledging their need for repentance of sins, the burial of their old way of life, and resurrection into new life. They are immersed in water, symbolizing death to self, and burst forth with life in Christ. With each movement of baptism, we draw closer to Jesus while fulfilling his will.

Finally, Jesus emphasizes teaching as a requisite for discipleship—specifically, teaching everything he has commanded. Proper teaching encourages others to grow in faith, becoming more like Christ. A life of sanctification involves repentance, submission, and commitment to following him. *Following* is a word picture of activity and intention.

Jesus taught through a combination of didactic teaching, Socratic conversation, and experiential learning. He preached to large crowds of people and smaller, more intimate gatherings. Often, Jesus spoke in parables revealing the truth of the kingdom of God. The largest portion of his time was spent living with, eating with, and discipling a small group of men who later changed the world.

The disciples were educated through their interactions with Jesus. By watching him preach, teach, and heal, they learned how to share the gospel, cast out demons, and treat the sick. Through these interactions, they realized the power of the Messiah. His ultimate sacrifice at Calvary affirmed their calling to carry forth his work to the ends of the earth. Their very education was accomplished "on the move."

Discipleship requires the greatest of sacrifices. It is the most significant investment of our time, energy, and resources. We pour out our souls into the lives of others. There are no shortcuts. Every move we make is with an eye on the prize, that others would come to faith through our witness. It can be accomplished only if we are ready and able to deliver the good news of Christ. For some it requires the stamina of marathon runners. For others it requires surrendering disability at the foot of the cross to glorify God.

QUESTIONS FOR REFLECTION

1. What is your normal level of physical exertion daily, weekly, monthly?
2. David proclaims that we are "fearfully and wonderfully made" (Ps. 139:14). What does this mean, and how were we designed to move?
3. Have you or a loved one experienced an injury or disability that impacts movement? What effect did this have physically, spiritually, and emotionally?
4. Running is a perfect metaphor for the Christian life of faith. It is unsurprising that Paul says, "I have fought the good fight, I have finished the race, I have kept the faith" (2 Tim. 4:7). In what ways are physical and spiritual races similar, different, and complementary?
5. In Matthew 28:18–20, Jesus says, "All authority in heaven and on earth has been given to me. Go therefore and make disciples of all nations, baptizing them in the name of the Father and of the Son and of the Holy Spirit, teaching them to observe all that I have commanded you. And behold, I am with you always, to the end of the age." Making disciples involves movement: going, baptizing, and teaching. How are you involved in each?

NOTES

1. Keh, A. (2019, October 14). Eliud Kipchoge breaks two-hour marathon barrier. *New York Times*. https://www.nytimes.com/2019/10/12/sports/eliud-kipchoge-marathon-record.html
2. Jones, A. M. et al. (2021). Physiological demands of running at 2-hour marathon race pace. *J Appl Physiol (1985) 130*(2), 369–79.
3. Centers for Disease Control and Prevention. (n.d.). Benefits of physical activity. Retrieved October 15, 2022, from https://www.cdc.gov/physicalactivity/basics/pa-health/index.htm
4. Marcdante, K. J., & Kliegman, R. (2019). *Nelson essentials of pediatrics*. Elsevier.
5. Carey, W. B. (2009). *Developmental-behavioral pediatrics*. Saunders/Elsevier.
6. World Health Organization. (2022). World report on disability. Retrieved October 15, 2022, from www.who.int/publications/i/item/9789240063600
7. Hogan, K. (2012, August 28). Athlete Aimee Mullins: Beauty in sports can be "liberating." *People*. https://people.com/style/athlete-aimee-mullins-beauty-in-sports-can-be-liberating
8. Mullins, A. (2010, December 26). A work in progress. *Moth Stories*. The Moth. https://themoth.org/stories/a-work-in-progress; Rosenbaum, M., & Zak, L. (2012, November 30). Aimee Mullins: Double amputee a model, athlete, inspiration. ABC. https://abcnews.go.com/US/aimee-mullins-double-amputee-model-athlete-inspiration/story?id=17851813
9. Bruce, F. F. (1988). *Book of Acts* (revised edition). New

International Commentary of the New Testament. Eerdmans.

10. Diman, A. et al. (2016). Nuclear respiratory factor 1 and endurance exercise promote human telomere transcription. *Sci Adv 2*(7), e1600031; Tucker, L. A. (2017). Physical activity and telomere length in U.S. men and women: An NHANES investigation. *Prev Med 100*, 145–51.
11. Sleiman, S. F. et al. (2016). Exercise promotes the expression of brain derived neurotrophic factor (BDNF) through the action of the ketone body beta-hydroxybutyrate. *Elife 5*.
12. Guure, C. B. et al. (2017). Impact of physical activity on cognitive decline, dementia, and its subtypes: Meta-analysis of prospective studies. *Biomed Res Int 2017.* doi: 10.1155/2017/9016924.
13. US Department of Health and Human Services. (2018). Physical activity guidelines for Americans. 2nd ed., https://health.gov/sites/default/files/2019-09/Physical_Activity_Guidelines_2nd_edition.pdf.
14. Gremaud, A. L. et al. (2018). Gamifying accelerometer use increases physical activity levels of sedentary office workers. *J Am Heart Assoc* 7(13).
15. Roberts, N. (2019, March 6). Americans sit more than anytime in history and it's literally killing us. *Forbes.* https://www.forbes.com/sites/nicolefisher/2019/03/06/americans-sit-more-than-anytime-in-history-and-its-literally-killing-us/?sh=1d127cc0779d
16. Joshua Project. (n.d.). Explore unreached peoples. Retrieved October 15, 2022, from https://joshuaproject.net

8

CREATE

In 1508, Michelangelo was commissioned by Pope Julius II to paint the ceiling of the Sistine Chapel. A budding sculptor, Michelangelo was yet unknown for his paintbrush. His rivals, the architect Bramante and the painter Raphael, hoped he would fail miserably. Instead, Michelangelo exceeded everyone's expectations.

What ensued rendered this once simple chapel for the cardinals into one of the most iconic spaces in the Vatican. As a masterpiece for the ages, the finished ceiling paintings came to epitomize the beauty of the High Renaissance period.

High Renaissance works are known for their realistic depictions of physical forms, use of light contrasting darkness, and linear perspective. Artwork from this short period between 1490 and 1530 emerged largely from Rome. Sculptures, architecture, and paintings typically focused on the majesty, glory, and beauty of the divine.[1]

Michelangelo used the technique of fresco to cover the entire ceiling of the Sistine Chapel. The Italian word *fresco*, meaning "fresh," describes the way the murals are painted on newly laid lime plaster. As water merges with colorful pigments into the plaster, the murals become part of the ceiling itself.

Painting on the ceiling of the Sistine Chapel required tremendous ingenuity and rigorous body control, because the highest arches were some sixty-five feet above the floor. Michelangelo himself designed the scaffolding that allowed him to create the frescoes in unconventional ways.

Forced to contort his neck at agonizing angles for almost four years, he developed severe neck pain, sore shoulders, and countless body cramps. Describing the gut-wrenching experience of painting, he wrote, "My beard toward Heaven, I feel the back of my brain upon my neck. My loins have penetrated to my paunch. . . . I'm not in a good place, and I'm no painter."[2]

Despite the obstacles, Michelangelo persevered, and the vaults of the Sistine Chapel have become one of the most recognized pieces of art worldwide. More than five million people visit every year, sometimes twenty thousand a day. They crane their necks upward to catch a glimpse of the masterpiece.

What they see are magnificent paintings of nine scenes from the book of Genesis. These include the conception of the world, the creation of man and woman, humanity's fall from grace, and the suffering that resulted. Two of the most extraordinary scenes on the ceiling are the *Creation of Adam* and the *Fall of Man*.

In the celebrated *Creation of Adam*, Michelangelo paints one of the most remarkable details of Christendom. Within this image, the finger of God reaches out to touch the hand of Adam. No one else would have imagined God's finger making contact with man's. Michelangelo's creativity has inspired countless reproductions of this painting, both metaphoric and exact. The painting depicts an omnipotent God who is neither distant nor detached but fully engaged in all of creation.

Andrew Graham-Dixon writes, "God writes on us with his finger. . . . The finger is the conduit through which God's intelligence, his ideas, and his morality seep into Man. And if you look at that painting very closely, you see that God isn't actually looking at Adam, he's looking at his own finger as if to channel his own instructions and thoughts through that finger."[3]

Michelangelo's masterpiece on the ceiling of the Sistine Chapel has inspired generations of artists, painters, and sculptors. His attention to detail while telling part of the redemptive story of Christianity yields an aura of reverence and worship. It points to the imagination of the divine and that of his creation.

THE GLORY OF GOD

God's creativity is revealed through general and special revelation. In Romans 1:20, Paul writes, "For his invisible attributes, namely, his eternal power and divine nature, have been clearly perceived, ever since the creation of the world, in the things that have been made. So they are without excuse." General revelation, or natural revelation, is the knowledge and understanding of God from natural phenomena. Special revelation, also called supernatural or direct revelation, is communicated directly to individuals or groups of people by God.

Michelangelo's first fresco depicts the creation of the heavens and the earth in Genesis 1:1. All of humanity can appreciate God's invisible attributes through the wonders of the works of his hands. The sky's bursting forth in light as the sun rises over the horizon provides evidence for God's eternal power and divine nature. As a symphony of color disappears over the clouds and into darkness, the end of one day marks the creation of another.

Whether through a beautiful sunset or towering mountains stretching into the clouds, God reveals his glory through the creativity of his workmanship. The psalmist confirms that "the heavens declare the glory of God, and the sky above proclaims his handiwork. Day to day pours out speech, and night to night reveals knowledge" (Ps. 19:1–2).

General revelation discloses common knowledge of the divine to widespread audiences. The tapestry of creation confirms God's existence. Witnessing his power, we are left without excuse to acknowledge his presence. Whether through our sight, smell, touch, taste, or hearing, our senses can only affirm his glory. According to Paul, what can be known about God is plain for all to see, because God has shown it to everyone (Rom. 1:19).

While God's existence is evident through his design, his plan of salvation is unveiled only through special revelation. God's ingenuity in creating this unique path to salvation is communicated directly to humanity by him. Before the Scriptures were available, this plan was revealed through visions and dreams, angels and prophets, and sometimes even directly spoken words.

Today, the gift of the Old and New Testaments reveals God's historical redemptive plan. The author of Hebrews bridges the gap between special revelation from antiquity and the new covenant: "Long ago, at many times and in many ways, God spoke to our fathers by the prophets, but in these last days he has spoken to us by his Son, whom he appointed the heir of all things, through whom also he created the world" (Heb. 1:1–2).

Not only is the creativity of God revealed through amazing tapestries of creation that surround us, but also it can be appreciated in the narrative of the Scriptures. Consider the story of the incarnation. By encountering the Word made flesh, we identify with Jesus' life, death, and resurrection. It's an ingenious way to communicate the message of the gospel to all nations.

When Jesus heals the blind, we see them open their eyes. As he teaches the disciples in parables, his words resonate with us in similar ways. As we hear James and John request to be seated at the right and left hand of the Messiah, we face our own selfishness and conceit. Jesus' final words at Golgotha—"My father, if it be possible, let this cup pass from me; nevertheless, not as I will, but as you will" (Matt. 26:39)—forever echo in our souls.

The contrast between the old and new covenants provides the backbone to our dealing with sin and redemption. The covenant of works is broken by humanity, whereas the covenant of grace is fulfilled by Jesus Christ. The creativity of this narrative allows every person to identify with their imperfections while considering their need for a savior.

Gordon Fee and Douglas Stuart attest to the power of the incarnation narrative, which invites us "vicariously to live through events and experiences rather than simply learning about the issues involved in those events and experiences."[4]

As we enter into the life of Jesus, and he enters into us by his Holy Spirit, we connect not only with his life but also with the truth of his teachings and the reality of his sacrifice. The life of the King becomes personal and real.

Though not commonly considered an attribute of God, creativity is an undeniable characteristic. It is no different from his more prominent attributes, which describe him as omnipotent, omniscient, omnipresent, immutable, infinite, and self-sufficient. Creativity is woven into the fabric of his being. Each of us inherits our imagination from God. Though we will never be as innovative as the one who formed us, our creativity reflects God.

THE SCIENCE OF CREATIVITY

It was a midsummer day in the heart of New York City. The air was hot, the humidity rising. My body grew restless as I sat in my office on the Lower East Side. With an invitation to speak at a conference in a few days, I was feverishly at work. Unfortunately, the words did not come easily. My thoughts were jumbled, and my mind was in disarray.

Instead of continuing to grind away at the talks, I decided to take a break and headed for the swimming pool to clear my head. After intermixing freestyle laps and the backstroke for half an hour, my arms and legs were spent. It was time to get out of the pool.

The comforting stream of water in the shower afterward refreshed my aching muscles. Before drying off, I made my way to the adjacent steam room, where warm vapor filled my pores. With my every breath, the soothing vapor slowly made its way into my lungs. My whole body felt at ease. A sense of tranquility and peace overwhelmed my soul.

After I'd drifted away for a few moments, unprompted words and thoughts filled my mind. Without thinking, I had an epiphany about each future talk. All of a sudden, it became clear what I needed to say and how the words should be conveyed. Through the inspiration of the Holy Spirit, specific passages of Scripture resonated within.

Fifteen minutes in the sauna turned out to be more effective than three full days of wracking my brain for the material.

Is it a coincidence that a breakthrough occurred in the steam room? Studies show that innovation and breakthroughs are more likely to occur when we're engaged in mundane tasks that require little thought. During these passive activities, our minds are better attuned to our stream of consciousness, or as Christians would say, the voice of the Holy Spirit.

Habitual activities such as taking a shower or sitting in the sauna allow our brains to enter into the default mode network.[5] DMN is our state of mind when not dynamically engaged in an activity. And fMRI (functional magnetic resonance imaging) shows that our brains light up in similar ways when at baseline or in a resting state. This default network takes a back seat during more challenging work, when the executive network helps make crucial decisions while focusing on a task.

When we are in the default mode network, we tend to ruminate on the past, consider the future, and reflect on our identity in light of those around us.[6] This perfect resting milieu encourages the formation of new thoughts and novel ideas. It's within this wandering state of mind that breakthroughs are made and epiphanies occur.

Spontaneous mind-wandering, compared with deliberate mind-wandering, enhances our ability to sift through information, thoughts, and ideas creatively.[7] In a study of writers and physicists, one-fifth reported that creativity occurred most frequently while engaged in spontaneous task-independent mind-wandering. In essence, they were engaging in an activity other than their chosen field of work and thinking about something unrelated to the newly discovered idea.[8]

But why do we sometimes find inspiration during mundane tasks while drawing a blank at other moments? It turns out creativity is extraordinarily complicated, and researchers are just beginning to understand it. While the default mode network plays a crucial role in creativity, collaboration with other networks may be just as important.

Research shows that a person's capacity to generate original ideas can be predicted from the strength of connectivity between the three major networks in the brain.[9] The default mode network may generate ideas, but the executive network evaluates them for feasibility. Meanwhile, the salience network determines whether these ideas are passed from the default network to the executive one. When all three networks work together, creativity is at its peak.

Genetically, some people are born with a greater gift of creativity. Compared with most, they have a special knack for bringing something novel into existence. Despite this, imagination can be nurtured even in those who struggle with inspiration. As children made in the image of a creative God, each of us possesses unlimited creative potential. Whether we extinguish our imagination or encourage its growth is up to us.

As cofounder of Pixar Animation Studios and president of Disney Animation, Ed Catmull fosters creativity as part of his job. For several decades, Pixar led the world of animation with hit movies like *Toy Story*, *Finding Nemo*, and *The Incredibles*. Each film was praised for its originality. Over time, Pixar became a pioneer in innovation.

While most companies' employees worked in individual office spaces, Pixar embraced large, open workspaces.[10] Workers' proximity to other people fostered natural collaborations. Employees were even encouraged to design and decorate their work areas to nurture artistic self-expression.

Initially, Pixar held meetings at long, rectangular tables with placards for specific individuals.[11] This unwittingly led to an environment of hierarchy and judgment. Over time, they moved to a square table without name cards. People at the fringes were no longer marginalized. An atmosphere of mutual respect for individual thoughts, ideas, and emotions led to vulnerability and candor and inspired creativity.

In this nurturing environment, freedom of expression flourished. With a sense of psychological safety, employees were free to fail. Most of the corporate world frowns on anything taking away from the bottom line, but at Pixar, failure was seen as the key ingredient to creativity. Catmull understood it as necessary for innovation.

By fostering risk-taking, Pixar engendered flexibility, ingenuity, and originality. They incorporated mistakes into the developmental phase of each film. By emphasizing the iterative process required for creativity, more time was allowed for the exploration and refinement of stories and ideas.

Pixar technology developers and engineers were given two days a month to pursue other captivating Pixar projects. Without the expectation of work and the pressure of deadlines, some of the finest contributions were made to projects unrelated to the employee's primary focus. Individual passions were often the driver of motivation.

Similar to the parable of talents (Matt. 24:14–30), each Pixar employee was gifted in different ways, and some multiplied their gifts more than others. By nurturing their interests, they were provided with every opportunity to do so.

SUBDUING OUR CREATIVITY

Sloths are well known for their endearing features. From furry round heads to buttonlike eyes, they rank among the most adorable animals. As tree-dwellers of Central and South America, these animals live in the canopies of tropical rainforests. Each day, they move forty yards at most within their habitat. Sloths can sleep up to twenty hours daily because of their remarkably low metabolism.

The term *slothfulness* is not commonly heard in today's vernacular. Synonyms like *idleness* and *laziness* are used more. But anyone who has seen a sloth in a nature documentary or at a zoo can get the picture of what slothfulness entails. Proverbs 19:15 teaches us that "slothfulness casts into a deep sleep, and an idle person will suffer hunger." The writer connects a lifestyle of laziness to the deep sleep that robs individuals of all their sensibilities. In this subdued state, people have no perception of reality. Their senses are diminished. This descent from passive idleness to debilitating slumber leads to hunger and even death (Prov. 21:25).

According to the Scriptures, being slothful is the antithesis of God's design for people. A reluctance to work is an affront to God. We were made to labor with our hands, heads, and hearts, to be creative in all aspects of our lives. Working and serving others gives dignity to our lives and relationships.

How do we reconcile the command not to be slothful

with the need for the spontaneous mind-wandering that promotes creativity? Today's fast-paced society values drive, productivity, and tireless effort. Being constantly wired leads to never-ending engagement with the world around us. Even our children are involved in every activity under the sun.

By not allowing our minds to entertain boredom, we suppress the creative juices needed to stimulate our imagination. When we don't permit our children to experience monotony, we rob them of the ability to fight it. It's often in these moments of creative idleness that innovation is born and stagnation is suppressed.

Creative idleness is purposeful freedom from the burdens and stress of daily life. Without freeing our minds to spontaneously (or deliberately) wander, we short-circuit the potential for discovery and invention. When the stress of expectations weighs on our souls, it's difficult to see the hidden miracles of life and perceive the voice of God.

While allowing some creative idleness in our lives engenders innovation, other behaviors can smother our imagination. With the advent of the internet and smartphones, the average time a person spends online per day has been steadily increasing since 2011. From 214 minutes per day in 2011 to 494 minutes per day in 2022, digital media has insidiously become all consuming.[12] Parents struggle with how much screen time to allow for their children. Fear of missing out leads young adults to constantly check their social-media accounts.

Our unrestricted access to information on the internet has changed the way we think, read, and learn. Some argue that unlimited access to knowledge hinders our creative genes. Since solutions to almost every problem can be watched on YouTube, people are less inclined to develop their own ideas. With answers to their assignments easily found on the web, students need only Google to finish their projects. Artificial intelligence programs like ChatGPT will even write a paper for you.

New content on social media is often targeted to receive likes, thumbs-ups, and shares. Freedom to fail and be creative is sometimes replaced by a desire to cater to the crowds. Quickly scrolling through Facebook, Instagram, and Twitter shows a common formula for positive and negative reinforcement. The most popular videos on YouTube follow the same format and delivery to optimize subscribers. One look at their titles and content tells us that content creators are sacrificing creativity for influence.

Yet there is a difference between watching mind-numbing videos and using the internet to enhance one's creativity. While spending hours chatting may not foster imagination, learning how to create a robot for a science project can inspire the next groundbreaking discovery. The rise of the internet age has revolutionized the potential for creativity. Today, digital technology grants access to knowledge and resources that once only the affluent could attain. An almost endless supply of teachers and content fosters learning and imagination.

Online communities provide support and encouragement for artistic expression, from painting to music to writing. How-to videos teach children to engage in every possible endeavor. They provide not only the basics but often also details and nuances taught by experts. Budding video producers can learn the ins and outs of their trade without leaving their homes. In the past, starving artists without the opportunity to sign a contract or the means to market their work were left in the dark. Now they have a venue to share their work. With one click of the mouse, the world can witness their artistry and judge their creativity.

Focusing on the quality of time spent online is more important than merely limiting the hours. In many ways, the internet allows us to develop our interests and grow our expertise in particular areas. Motivation originates from different places. Intrinsic motivation reflects a person's passions and interests, while extrinsic motivation comes from outside rewards or threats. Not surprisingly, intrinsic motivation has a far greater impact on our creativity.

The rapidly expanding field of artificial intelligence may threaten our innovation and motivation. Artificial intelligence creates work based on what exists. It analyzes knowledge, data, and patterns to predict outcomes in the future. Every AI output is based in part on past information. By relying solely on artificial intelligence, we risk sacrificing the freedom to create.

Teresa Amabile argues that expertise and creative thinking are the raw materials for innovation.[13] But intrinsic motivation determines whether people possess the drive to follow through. A biomedical engineer whose children suffer from juvenile rheumatoid arthritis is more likely to create novel prosthetic devices for those with arthritis than one without a direct connection to the disease.

The perfect balance of personal expertise, creative idleness, and intrinsic motivation is needed to provide inspiration and tenacity. Behaviors that disrupt each of these areas can derail our imagination. By limiting these negative influences, we are less likely to follow the path of slothfulness described in Proverbs. And we are more likely to remain open to the exceptionally creative thoughts that originate with God, are communicated by the Holy Spirit, and are often left little room to operate if our minds and imagination are flooded with the nonsensical, unhealthy, or wasteful.

REFLECTING THE GLORY OF GOD

As the capstone assignment for a class I teach at Alliance Theological Seminary, students are required to design an original gospel presentation for evangelistic purposes. The presentation is to be useful beyond seminary training. Grading is based on theological grounding, cultural relevance, fitness for future use, and creativity.

With flexible guidelines, students are encouraged to use their imagination to make something out of the ordinary. Because it is the last assignment of the semester, they have ample time to incubate, innovate, and create. Fifty percent of the grade comes from their peers.

Most projects far exceed everyone's expectations. Regardless of innate artistic ability, everyone is amazed by the creativity of other students. Each project tells the story of salvation in a unique way. Spoken words target culturally relevant people groups. Animation creates images that stir the imagination of nonbelievers. Movement through dance touches the heart more than voices from worship music. Social-media posts reach segments of the population not easily accessed by traditional approaches.

One future missionary designed a comic book targeting families from minority backgrounds. Carefully narrated captions revealed the truth of the gospel in innovative ways.

He transformed his own children into cartoon figures, making the teachings of the Bible come alive for other kids.

An intercultural studies student created a 3-D gospel origami finger game. By adapting the timeless origami finger game to a framework of guilt-innocence, shame-honor, and fear-power, players identified the circumstance, worldview, and culture necessary for gospel penetration. Different questions were coupled with carefully designed figures on the triangular pieces to eventually reveal each player's cultural preference.

A fitness guru with a lifelong passion for exercise developed a Christian workout video combining traditional aerobic exercises with targeted scriptural references and complementary music to provide a sensory experience. By prompting movement of the arms and legs, stimulating the mind, and stirring the heart, the video helped engage every part of participants' bodies in worship.

When we create, we emulate the Creator.

Ephesians 2:10 says, "We are God's handiwork, created in Christ Jesus to do good works, which God prepared in advance for us to do" (NIV). The Greek word for handiwork, *poeima*, is the origin of the word *poetry*. Historical readers of Paul's writing understood the artistic ramifications of poetry for God's creation. Not only are we made with the hands of God, but we are his tapestry, his poem, his artistic expression, his masterpiece.

As sentient beings, we can recognize and appreciate God's creativity. Through our eyes, we witness the beauty of snow-capped mountains towering over the horizon. The sound of cascading waterfalls reverberates through our ears as droplets bounce off the weathered rocks below. Our taste buds identify sweet, salty, and savory while enjoying our favorite culinary delights.

When we appreciate God's workmanship, it is our choice whether to respond in praise. Creativity is a natural outflowing of this worship. Formed in the image of God, humanity alone is able to make poetry of all things. Creativity fills previously blank spaces devoid of expression with radiance from our imagination. Whether it's stimulating conversation during fellowship, written words describing our encounters with God, or the artistic use of our hands in worship, we are creating. It's what God designed us to do.

In Genesis 1:28, God encourages Adam and Eve to "be fruitful and multiply and fill the earth." Our bodies were made to create through innovation and imagination, artistic gifts and talents. But they are also designed to procreate, fill the earth, and subdue it. When we do so, we fulfill God's mandate and receive his blessing.

Though not everyone will experience the ability to give birth and raise children, those who do have the opportunity to create and be creative. Proverbs 22:6 exhorts parents to "train up a child in the way he should go; even when he is old he will not depart from it." Teaching our children the discipline and love of the Lord requires every ounce of creativity in our bodies. Innovation is requisite in our planning. Our approach to discipline must include careful discernment.

Every day presents new challenges. Temper tantrums derail planning. Unforeseen circumstances steal away joy. Sudden illnesses threaten our faith. Yet we have tremendous opportunities to nurture, love, and disciple our sons and daughters. When we do so with creativity, our biological and spiritual children become like "arrows in the hand of a warrior" (Ps. 127:4).

INSPIRED BY THE SPIRIT

By embracing creativity in our lives, we worship incarnationally. Just as the Word was made flesh to pave the way for our salvation, so we have been filled with the Holy Spirit for our bodies to exalt him. When we create, we affirm the natural revelation of God while reflecting his glory. Every part of our bodies honors God as we become instruments of his artistry.

Modern science tells us that the recipe for imagination involves creative idleness, subject-matter expertise, and intrinsic motivation. Yet the key ingredient missing from this secular formula is inspiration from the Holy Spirit. From this stirring, the creativity of broken people can be transformed into something new with significant potential for kingdom impact.

With the inspiration of the Holy Spirit, our creativity has a purpose beyond ourselves. It becomes aligned with God's will. When we receive the grace of God through faith, we become a new creation in Christ. The old has gone and the new has come (2 Cor. 5:17). With the power of the Holy Spirit, our lives embody the good works described in Ephesians 2:10. This allows our creativity to stir the hearts and minds of others toward Jesus.

Secular creativity inspires others with a sense of awe for the artist. Holy Spirit–filled creativity stirs others to see the face of God and worship him. The imagination to paint, build, draw, sew, write, or worship may look similar for both, but only one points to the author of life and gives him glory, honor, and praise.

Holy Spirit–inspired creativity is a selfless act. It not only points to God but also serves our neighbors. The Spirit of

God convicts us to love others in fresh new ways. He softens our hearts toward the outcast, vulnerable, and unlovable.

When we allow God to mold our lives for his purposes, we become his intended handiwork. The more we create, the more we see the poetry of God all around us. By using the raw ingredients that God provided, we appreciate that God himself is part of everything we make.

During the summer of 1741, composer George Frideric Handel wrote the *Messiah* in a little more than three weeks. He did so alone with the inspiration of the Holy Spirit. A servant once overheard Handel proclaim, "I did think I did see all heaven before me and the great God himself."

Since the first performance a year later, the *Messiah* has become synonymous with Christmas. Non-Christians and believers alike crowd concert halls to hear the symphonic melodies. Breathtaking oratorios are sung by soloists as the choir sings in harmony. Uplifting messages on the life, passion, and resurrection of Christ complement the sound of voices and instruments. They tell a story of a creative God who sent his Son to walk and run, thirst and hunger, laugh and weep, teach and listen to us. Through his incarnation, Jesus triumphs over sin and death. Everything we create, whether music, art, photography, or food, can point to this savior.

On May 22, 2021, New York Chinese Alliance Church partnered with muralist Bianca Romero, Christian arts nonprofit THRIVE Collective, and the City of New York to unveil a three-story, gospel-centric, Asian-American Pacific-Islander (AAPI) mural on the wall of our church facing Delancey Street.

We shared the gospel with countless neighbors on the Lower East Side, a diverse community that includes Asian, Black, White, and Latino populations. More than thirty tables represented a plethora of NYC agencies, such as the Office of Immigrant Affairs, Commission on Human Rights, Department of Education, Office for the Prevention of Hate Crimes, and Housing Authority, providing resources to the outcast and vulnerable. Numerous Lower East Side restaurants served food, while the Department of Health offered free vaccinations and other services. Christian DJs delivered a block-party atmosphere, while a multiethnic jazz band lifted everyone's souls.

The mural depicts three Asian women, one elderly grandfather, and a little boy. Each is holding a flower as a sign of solidarity and peace. Their faces say to the downcast, vulnerable, and oppressed, "You are beloved, safe, and welcome here." Underneath their silhouettes lies the answer to the question of Micah 6:8, "What does the LORD require of you?" The answer? "To act justly and to love mercy and to walk humbly with your God" (NIV).

Through the creativity of a muralist, the speechless are given breath. The joyless experience joy. The brokenhearted receive hope. And the Word of God reverberates as an encouragement of truth.

QUESTIONS FOR REFLECTION

1. In what ways can you witness God's creativity through general and special revelation?
2. How will the science of creativity change the way you approach innovation?
3. How can you reconcile the command not to be slothful with the need for spontaneous mind-wandering that promotes creativity?

4. How can technological advances subdue your creativity? In what ways can the internet foster innovation?
5. Ephesians 2:10 says, "We are God's handiwork, created in Christ Jesus to do good works, which God prepared in advance for us to do" (NIV). In what ways can you foster creativity in worshiping God? How can you share the gospel in creative ways?

NOTES

1. Forbes, M. (2020, May 10). The Sistine Chapel: Unfolded and explained. The Collector. https://www.thecollector.com/sistine-chapel
2. Graham-Dixon, A., & Michelangelo, B. (2009). *Michelangelo and the Sistine Chapel.* Skyhorse.
3. Graham-Dixon, A., & Michelangelo, B. (2009). *Michelangelo and the Sistine Chapel.* Skyhorse.
4. Fee, G. D., & Stuart, D. K. (2014). *How to read the Bible for all its worth.* Zondervan.
5. Raichle, M. E. et al. (2001). A default mode of brain function. *Proc Natl Acad Sci USA 98*(2), 676–82; Raichle, M. E. (2015). The brain's default mode network. *Annu Rev Neurosci 38*, 433–47.
6. Buckner, R. L. (2013). The brain's default network: Origins and implications for the study of psychosis. *Dialogues Clin Neurosci 15*(3), 351–58.
7. Mooneyham, B. W. et al. (2017). States of mind: Characterizing the neural bases of focus and mind-wandering through dynamic functional connectivity. *J Cogn Neurosci 29*(3), 495–506.
8. Gable, S. L. et al. (2019). When the muses strike: Creative

ideas of physicists and writers routinely occur during mind wandering. *Psychol Sci 30*(3), 396–404.

9. Beaty, R. E. et al. (2018). Robust prediction of individual creative ability from brain functional connectivity. *Proc Natl Acad Sci USA 115*(5), 1087–92.
10. Catmull, E. E., & Wallace, A. (2014). *Creativity, Inc.: Overcoming the unseen forces that stand in the way of true inspiration.* Random House.
11. Catmull, E. E., & Wallace, A. (2014). *Creativity, Inc.: Overcoming the unseen forces that stand in the way of true inspiration.* Random House.
12. Statista. (2022). Average time spent per day with digital media in the United States from 2011 to 2024. Retrieved November 9, 2022, from https://www.statista.com/statistics/262340/daily-time-spent-with-digital-media-according-to-us-consumsers
13. Amabile, T. (1998, September–October). How to kill creativity. *Harvard Business Review.* https://hbr.org/1998/09/how-to-kill-creativity

9

REST

It was a hot, humid summer day in the heart of the South. As I walked out of the hospital with bags under my eyes, sweat oozed from every pore. A grueling twenty-four-hour overnight shift was done at last. From the time I stepped onto the ward until signing out that next day, a black cloud seemed to follow me everywhere.

The popular saying "When it rains, it pours" had taken on a whole new meaning the previous night. Every patient I cared for required special intervention, more than expected. Clinical supervision is typically intuitive, but nothing was straightforward that evening. The presenting signs and symptoms of the patients did not lead to simple diagnoses. To ensure nothing was missed, diagnostic plans were necessarily complex.

Compounding my clinical haze, my pager kept beeping every few minutes. From the first minute to the last hour, I could barely keep up with the growing number of calls for my attention. Emergency room physicians paged to admit new patients. Nurses requested orders and medication. Family members yearned for updates on their loved ones. There was no time to use the bathroom or stay hydrated.

But, as you know, time flies when you're busy. In the blink of an eye, twenty-four hours flew by. After signing out to the next provider, I experienced a strange sense of euphoria. I was never more delighted to turn off a pager. Though my body was weary, my mind felt liberated.

With two full days off, I planned to drive home to Atlanta for the weekend. After being away from family and friends for several months, the thought of a home-cooked meal was irresistible. Would a short nap be wise before embarking on the two-and-a-half-hour drive home? Considering how long I had slept after the last on-call shift, there was a distinct possibility I'd never make it out of town.

Determined to see my family and loved ones, I packed my bags and stocked up on caffeinated beverages. I cranked up the air conditioner and set the radio to "blaring" to keep me alert. Then I pressed the gas pedal and started the 158-mile journey.

The first hour went smoothly. As the sun beat down on my silver Ford Probe, I was glad to watch the hash marks tick by and sip an ice-cold Coke. Adrenaline from the previous night's work and euphoria from being off kept me awake and alert.

During the second hour, acute sleep deprivation caught up. Caffeine, sugar, and music no longer curbed the overwhelming need for sleep. In hindsight, I should have listened to my body's warning signs and stopped at a rest area.

Instead, I continued to drive. With each passing minute, I fended off the escalating urge to close my eyes. Attempts to sing at the top of my lungs to the radio tunes were futile.

There was no way to stay awake. No hope to fight the exhaustion.

Suddenly, I found myself speeding along the grassy

median dividing traffic. Startled, I jerked the car back onto the highway. Unfortunately, this caused the car to skid across the asphalt straight into the trees on the side of the road.

My life flashed before my eyes. I thought, *This is surely the end.*

Miraculously, God preserved my life. Instead of getting into a head-on collision, my car ended up sandwiched between two large oak trees. The airbag deployed in my face as the seat belt kept me from flying through the windshield.

Highway I-20 is one of the most dangerous freeways in Georgia. It has been dubbed "Die-20" because of the high rate of fatalities. The freeway runs west to east, connecting Alabama to the Savannah River. Because of its remarkably straight path, cars can easily reach speeds above the posted seventy-miles-per-hour limit.

On this day, the incident wasn't caused by my speeding along the interstate. Instead, it was because of my poor choices after a night of acute sleep deprivation.

One in five car crashes occurs because a driver is groggy.[1] People who sleep five or six hours a night are twice as likely to crash as those who sleep more than seven hours.[2] Unfortunately, this trend increases proportionally to the lack of sleep an individual experiences. For those who habitually get only four to five hours, the crash rate is equivalent to that of a drunk driver.

THE SCIENCE OF REST

"And he said to them, 'Come away by yourselves to a desolate place and rest a while.' For many were coming and going, and they had no leisure even to eat" (Mark 6:31).

When Jesus spoke these words to the disciples, he knew they were exhausted. After being sent off two by two to preach repentance, they began casting out demons and healing the sick. They witnessed Jesus' rejection at Nazareth and John the Baptist's brutal murder by Herod. Now they faced a hungry crowd of five thousand.

Jesus knew how important it was for the apostles to rest. Recognizing the physical, mental, and spiritual toll ministry had taken, he had compassion on them. Their bodies, minds, and souls needed time to recuperate.

Few doubt that rest is beneficial. A good night's sleep makes us feel refreshed and revitalized. We have more energy to move about and even to exercise. Our focus improves and our engagement with others is heightened.

On the other hand, lack of sleep can throw our whole lives out of sync, affecting work, stamina, relationships, and health.

What happens when we rest? After years of study, researchers still don't have a full understanding of sleep. Yet they have uncovered much in the last few decades.

The structure of our slumber is described as sleep architecture. It comprises the stages and cycles of an individual's sleep. We undergo multiple sleep cycles while resting, with sleep stages lasting up to 120 minutes. Good-quality rest is the result of healthy sleep architecture.

Almost immediately after someone falls asleep, the body undergoes significant changes. Vital signs like heart rate, breathing, and temperature decrease. While there is little movement outwardly, the body experiences multiple sleep cycles while resting.

Four major stages of sleep are divided by whether rapid eye movement (REM) occurs. Stages 1 through 3 (non-REM) of sleep grow increasingly longer, from one to five minutes for stage 1 to twenty to forty minutes for stage 3. REM sleep lasts from ten to sixty minutes.[3]

Each stage of non-REM sleep involves decreasing brain activity and increasing relaxation. Not surprisingly, it's much easier to be aroused during the earlier stages of sleep. As the last stage of non-REM sleep, stage 3 contributes to memory, thought, and renewal of the body. Tissues, bones, and muscles are repaired while cells are restored.

During stage 4, appropriately named rapid-eye-movement sleep (REM), activity in the brain crescendos. The eyes randomly flicker without sending visual information to the brain. Meanwhile, the rest of the body faces temporary paralysis with low tone. REM sleep is crucial for learning, memory, and the production of proteins. It is within REM sleep that the most intense dreaming occurs.

As I was driving down I-20, my body felt an uncontrollable urge to nap. No matter how hard I fought, falling asleep was inevitable. Every extra hour I stayed awake built up the pressure to lie down.

The inertia from our homeostatic sleep system is one of the two methods the body uses to regulate sleep. During the day, a person's need for rest progressively increases. Normally, the greatest pressure occurs just before bedtime. Periods of insufficient rest cause us to sleep more deeply and for longer durations.

The circadian rhythm, our body's natural biological clock, is the other regulator of sleep. Circadian rhythms average twenty-four hours, the length of a normal day. After a good night's rest, the circadian waking drive increases, causing individuals to rise and shine. Typically, this wakefulness is promoted during the day and declines in the evening, coinciding with natural sunlight. Synchronization of the circadian system with light-dark cycles of sunlight encourages sleepiness at night and wakefulness during the daytime.

Learning and memory are highly dependent on healthy sleep. Though the acquisition of new information and the recall of previously stored memories occur during wakefulness, consolidation of memories transpires during sleep.[4] As individuals pass through the later stages of sleep, neural synapses vital for recall are created and strengthened.

A good night's rest can heighten one's emotional intelligence by helping people process emotional cues around them. Not only does sleep help synthesize these signals, but it even increases empathy toward others.[5] Healthy sleep makes us more patient, kind, and understanding with our spouses, children, and coworkers.

Proper sleep is vital to the physical and mental development of adolescents, children, and newborns.[6] Later in life, it sharpens the brain while clearing harmful chemicals accumulated over time.[7] Many of these toxins lead to neurodegenerative illnesses such as Alzheimer's disease.[8]

Healthy sleep also promotes healing, growth, and regeneration of tissues by releasing growth hormones. During sleep, many hormones are secreted that affect appetite, metabolism, and sugar processing.[9] People who sleep eight hours or more have a lower body mass index than those who sleep less than six hours per night. Adequate rest even boosts the immune system, helping to fight against infectious diseases.[10]

Sleep is one of our most effective medications. No pill comes close. Healthy rest treats a plethora of ailments and prevents even more. And unlike most prescriptions, sleep has no side effects and comes with no financial cost.

Since sleep impacts almost every bodily system, we can understand why sleep deprivation is detrimental. Sleep is truly a wonder drug. It doesn't serve just one biological purpose but fulfills a plethora of roles within the body. With its so many proven health benefits, it's a wonder more people don't find adequate rest.

It turns out *we* are our own worst sleep saboteurs.

RUINING OUR SLEEP

One summer morning, a college student, John, visited my pediatric practice. His medical record was pristine. He had no significant medical history. He was not taking any medications. His vital signs were stable, and his weight was excellent. The rest of the physical exam was normal. By all accounts, he was strong, rigorous, and healthy.

John's concern was not outwardly evident. He struggled with insomnia. I asked him to explain. As a college student, he stayed up all night, sometimes completing homework and assignments but mostly playing video games with other night owls. Often he would not go to bed until sunrise and slept late into the afternoon.

As a naturally gifted student, John had no problem keeping up with his grades his freshman year. Because of his aptitude, he was able to skip classes during the day and catch up with the workload at night. Introductory courses typically have large numbers of students, so the professors never noticed his frequent absences.

But during his sophomore year, John was finding it increasingly difficult to keep up with the coursework. Higher level classes were smaller and more intimate. Interaction with others had become crucial for his learning. Attendance was often part of the grade.

John's parents prodded him to move away from this nocturnal lifestyle. They tried forcing him to rearrange his schedule and stop playing video games. It worked for a while. But like a bungee cord, he always bounced back to his old pattern.

I diagnosed John with delayed sleep phase syndrome. Compared with the general population, it is quite common in adolescents and college students.[11] If he had struggled with this problem since early childhood, a referral to a sleep disorders specialist would have been warranted. But for most college students, the diagnosis is a transient phenomenon instead of a permanent problem.

Sleep is designed to be restful, restorative, rejuvenating. In *Macbeth*, Shakespeare describes it this way: "Innocent sleep. Sleep that soothes away all our worries. Sleep that puts each day to rest. Sleep that relieves the weary laborer and heals hurt minds. Sleep, the main course in life's feast, and the most nourishing." Unfortunately, poor sleep hygiene is one of the major culprits of insomnia. Like John, many people engage in activities and hobbies that keep them up late. Artists, writers, and musicians are especially prone as they immerse themselves in their passions.

But poor sleep hygiene doesn't affect creative types only. It is nearly endemic in today's fast-paced, permanently online world. Working remotely often means being online or checking email 24/7. Catching up on social media before bed can easily delay sleep for several hours. Parents of young children are all too aware of the sleep issues their kids face.

With so much extraneous noise, poor sleep hygiene

becomes a vicious cycle. Remember that homeostatic sleep drive? With all of the factors I've mentioned, pretty soon, there is no longer enough pressure to fall asleep early, even when desired. It's little wonder that so many people are prone to late nights.

Compounding nocturnal tendencies to stay up late is the propensity to sleep in later, especially on the weekends. A hard week of work or school can lead to a paucity of sleep and the yearning to catch up on lost rest. Unfortunately, artificially changing the time we wake up significantly impacts our ability to fall asleep later that night. So do caffeinated drinks like soda, coffee, and tea.

We were created to rest, but the challenges are real. Stress from relationships and work permeates our lives. Endless distractions from smartphones and computers abound. Environmental and political concerns, both local and systemic, surround us.

Whether positive or negative, our behaviors have an enormous impact on the quality and quantity of our sleep. Over time, sleep patterns become normalized, for better or worse.

When I was working in the field of global health, I traveled more than 40 percent of the time. Some countries were a few hours behind, others a half day ahead. Jet lag always hit hardest the first twenty-four hours. But over time, I realized how to quickly adjust my circadian rhythm to the new location and time.

If I fought the urge to sleep and stayed awake the first day, my body would be exhausted and ready for bed that evening. Settling into a normal work, exercise, and meal routine as soon as possible was paramount. If I stuck to my regular sleep and wake cycle while eating, resting, and working in the patterns I normally used at home, my body soon adjusted to the new location and time zone.

The key to protecting rest is a pattern of predictability. Our bodies crave discipline and regular schedules. That wisdom extends beyond daily sleep to weekly rest.

SABBATH AND PATTERN OF RESTS

Since the creation of the world, God ordained a pattern of weekly rest for all humanity. Exodus 20:8–11 describes this day: "Remember the Sabbath day, to keep it holy. Six days you shall labor, and do all your work, but the seventh day is a Sabbath to the LORD your God. On it you shall not do any work, you, or your son, or your daughter, your male servant, or your female servant, or your livestock, or the sojourner who is within your gates. For in six days the LORD made heaven and earth, the sea, and all that is in them, and rested on the seventh day. Therefore the LORD blessed the Sabbath day and made it holy."

The Hebrew word *shabbat* means "rest." It occurs 111 times in the Old Testament, with more than a third of its references found in the Pentateuch. The earliest example of Sabbath is seen in Genesis, where God models the holy pattern of weekly rest he intends for humanity.

After six incredibly productive days, God rests. He has just created the heavens and earth and put the stars in their places. After allowing the land to sprout with vegetation and living creatures to inhabit the surroundings, he makes man and woman in his image.

An omnipotent, omniscient, and omnipresent being

doesn't need a break. Yet he takes one anyway so we would follow suit. With the creation of Sabbath, God ordains a pattern of rest that is vital for all. It affects not only our physical health but our spiritual well-being.

In Genesis 2:3, God blesses the Sabbath and makes it holy. According to Exodus 31:16–17, it was a sign between God and his people. Sabbath reflects God's covenant and his desire to be in communion with believers. Every seventh day of the week, we worship *and* rest.

By emphasizing the sanctity of rest, God establishes a connection between work and repose. The fourth commandment confirms that for six days we are to labor, but on the seventh day we rest—just as God did.

In all things, God establishes a rhythm of labor and rest for us. This natural rhythm provides respite from the toil of our daily lives. Yet it also invites us to faith and intimacy with God.

Patterns of sleep for all of creation were built into the genesis of the world and its creatures. Day and night, not endless day. Night followed by dawn, not endless night. Even seasons of dormancy for plants and animals were important elements of creation.

Rest begins with nightly sleep that allows us to feel refreshed and rejuvenated for each new day. The benefits of healthy sleep are numerous, as we've seen. Yet choosing to sleep is also a remarkable demonstration of faith. It shows our recognition of Christ-in-us "holding down the fort" or remaining on alert on our behalf even while we sleep.

David confirms this when he says, "In peace I will both lie down and sleep; for you alone, O Lord, make me dwell in safety" (Ps. 4:8). Though thousands of enemies surround him, David confesses that it is God alone who can sustain him.

From patterns of daily sleep, we move to disciplines of weekly rest. In a world where connectivity is king, it has never been harder to honor the Sabbath. There's an expectation, especially with lawyers, military members, moms, and healthcare providers, of being "on call" twenty-four hours a day, seven days a week. Yet studies have repeatedly reinforced the wisdom of Sabbath.

Regularly scheduled time off leads to better productivity, more open dialogue, and new ways to work efficiently and effectively.[12] Sabbath-keeping is associated with improved quality of life, increased satisfaction with relationships and family, and better spiritual health.[13]

Beyond daily and weekly patterns of rest, the Bible presents another level. The Israelites would pilgrimage to Jerusalem several times a year. During these special occasions, they left everything behind to gather in the house of the Lord. Living in an agrarian society, the Hebrew people were busy with their families, working hard to survive, farming and tilling the land, raising their livestock and watching over their sheep.

Then it was time to go to the house of the Lord. Time to worship. Time to sacrifice. Time to feast. Children, adults, and elderly, the Israelites rested. No responsibility superseded the obligation to visit the temple.

Some came from far away; others lived near Jerusalem. All gathered in the temple of God with the people of God. Together, the Israelites rejoiced, remembering what the Lord had done for them. Whether commemorating the Passover, the Festival of Tabernacles, or the Feast of Harvest, they were God's chosen people, anointed as a holy nation, and, leaving behind their day-to-day responsibilities, they met to acknowledge that fact.

A pattern of rest was so vital to God that even the fields were provided opportunities to take a break. Leviticus 25:1–7 speaks of the Sabbath year when the lands lay fallow. As a common agricultural practice, cessation of farming allowed the soil to rest and prevented diseases from infecting the land. Fruits and vegetables that grew naturally during the sabbatical year were made freely available to all, teaching all to acknowledge and depend on God.

Daily, weekly, and periodic times of respite are all patterns of rest engineered to encourage and bless us. Jesus himself confirms that "the Sabbath was made for man, not man for the Sabbath" (Mark 2:27). With the proper balance of labor and rest, we are able to more fully engage in the work of the kingdom while entering into communion with God and enjoying his creation.

THE HEART OF SABBATH REST

In the Old Testament, a second Hebrew word, *menuhah*, describes rest. For the Jewish people, it means far more than an absence of work. *Menuhah* is described as tranquility, peace, and harmony. Yet it is also intertwined with the concept of holiness.

Jewish scholar Abrahm Heschel argues that *menuhah* was introduced on the seventh day. Creation wasn't complete until God made it. In defining an architecture of holiness, specific times were ordained for communion with God. Sabbath provides these opportunities to enter into physical and spiritual rest.

Honoring the Sabbath, we find healing for our bodies and nourishment for our souls. Through the design of time, God created a venue for rest on the Sabbath. There we experience *menuhah*.

In the Old Testament, every seventh day was earmarked for this rest. But the incarnation of Jesus Christ makes Sabbath rest available at all times. Jesus says, "Come to me, all who labor and are heavy laden, and I will give you rest. Take my yoke upon you, and learn from me, for I am gentle and lowly in heart, and you will find rest for your souls" (Matt. 11:28–29).

The commandment to keep the Sabbath holy is designed to forge rest among God's people. Yet it is challenging to find repose when we are slaves to the law. Though the Jewish people can choose whether to obey the fourth commandment, the law itself cannot turn their hearts to God. They can observe the Sabbath, but it will not necessarily draw them closer to the Lord.

Mark 2:28 tells us that Jesus is Lord, even of the Sabbath. What the Sabbath law cannot accomplish, Jesus fulfills through his incarnation. Through his life, death, and resurrection, Jesus frees us from the condemnation of the law (Rom. 8:1–4). He releases us from the works of the law while inviting us to rest in his work of redemption. By his sacrifice on the cross, we are forgiven of sin and freed to enter into Sabbath rest. The writer of Hebrews describes this Sabbath rest for the people of God: "Whoever has entered God's rest has also rested from his works as God did from his" (Heb. 4:10).

Sabbath rest transcends space and time. It is available to believers regardless of the hour or day. The promise of rest begins with the confession of faith from our mouths and within our hearts (Rom. 10:9–10). In the crucified and resurrected Christ, Sabbath rest is available to all humanity and communion with God is restored.

We will experience the perfection, completeness, and fullness of Sabbath rest only in heaven. On earth, when it is observed as it was designed, Sabbath is a sweet foretaste of eternity, where a great multitude from every tribe, tongue, and nation will stand before the throne singing, "Salvation belongs to our God who sits on the throne, and to the Lamb!" (Rev. 7:10).

Jesus inaugurated Sabbath rest for this life, but it will not be consummated until eternity. Though we have access now, it is our choice whether to embrace it.

DISCIPLINES OF REST

Hebrews 4:11 exhorts us to "enter that rest, so that no one may fall by the same sort of disobedience." Though this passage speaks to eternal rest, we were created to rest daily, weekly, periodically. When Jesus sent the Holy Spirit to indwell our hearts, our bodies became a resting place for God. As temples of the Holy Spirit, we honor God by resting spiritually, mentally, and physically.

Ironically, rest requires work. Just as God was devoted to rest on the seventh day, we too commit to timely and scheduled rest. By purposely setting aside an entire day after creation, he affirmed the temporary cessation of work in our lives.

Whether it's nightly sleep, weekly Sabbath, or periodic time away, disciplines of rest require devotion. Our ability to cultivate these habits that are ingrained by commitment depends on our devotion and perseverance. Sometimes we succeed and sometimes we fail.

In 1 Timothy 4:7 (NASB), Paul teaches us to "discipline [ourselves] for the purpose of godliness." The goal of Sabbath is thus sanctification—making us more like Christ—in addition to physical rejuvenation and communion with God. We aren't godly because we embrace Sabbath, we *become* godly by observing the Sabbath. Misunderstanding that was the Pharisees' cardinal mistake.

In reality, all three goals—sanctification, rejuvenation, and communion with God—should be the focus of rest. How wonderful it would be if we viewed every opportunity for rest within this frame. During moments of Sabbath rest, we can be devoted to the apostles' teaching and fellowship, to the breaking of bread and prayers (Acts 2:42). Similar to the early church, we can choose to worship, study the Word of God, or commune with the body of Christ.

What about sleep? We know it provides physical healing and restoration of our bodies, but how can we foster sanctification and communion with God while not awake? The same way we tie spiritual disciplines to Sabbath or times of periodic rest.

As we engage our hearts and minds in the daily practice of spiritual disciplines, sleep consolidates our learning and solidifies our memories. Researchers have shown that sleep before learning helps prepare our brains for the initial formation of memories, while sleep after learning is essential to saving new information within the brain's neural connections.[14] Both the night before and the night after memorizing the Word of God are important to hardwiring these truths in our minds long term. Similarly, learning a new praise song or integrating biblical truths requires good sleep.

What's more, it's most common to dream about things we're already thinking about. Perhaps that's why we should meditate more on the Word of God before bed instead of rehashing the worries of the world. When we tie traditional spiritual disciplines to daily patterns of sleep, we train ourselves for godliness while enhancing our communion with Christ.

No matter the frequency, rest is a gift from above, both now and for eternity. As created beings, we need to rest. We need to trust that God is at work when we are not. We need to acknowledge that he alone is God. We are not.

Disciplines of rest remind us to worship the triune God while taking a break from our work. They invite us to embrace our limitations as created beings. They teach us to relinquish worries and fears while yielding control of our lives.

Embracing rest, we confront the idolatries of productivity and agency so prevalent today. When we do so, we affirm God's transcendence and indwelling. We proclaim our trust in him instead of ourselves.

When we rest, we step out of the driver's seat and allow God to take the wheel. We give up our sense of responsibility to be in charge, whether at home, at work, or in the family. Human nature says we can't afford to rest. Biblical teaching tells us we can't afford not to.

When we rest, we acknowledge that we are expendable. We are not the saviors of our jobs, our children, our congregations, or our clients.

Only one savior exists. His name is Emmanuel.

QUESTIONS FOR REFLECTION

1. We are our own worst sleep saboteurs. In what ways do you sabotage your ability to rest?
2. The science behind rest teaches us the benefits of a well-rested body along with the consequences of one that is sleep deprived. Which have you experienced?
3. The Bible describes patterns of daily, weekly, and periodic rest. How can you better engage in these disciplines of rest?
4. The writer of Hebrews encourages us to enter into Sabbath rest. How is this different from the patterns of rest we've described?
5. How can you integrate traditional spiritual disciplines with necessary patterns of rest?

NOTES

1. National Sleep Foundation. (2022). Sleep in America poll. Retrieved November 19, 2022, from https://www.thensf.org/sleep-in-america-polls
2. Vaughn, B. V., Stutts, J. C., & Wilkins, J. W. (1999). Why do people have drowsy driving crashes? Retrieved November 19, 2022, https://rosap.ntl.bts.gov/view/dot/40418.
3. Colten, H. R., & Altevogt, B. M. (eds.). (2006). *Sleep disorders and sleep deprivation: An unmet public health problem*. National Academies Press.
4. Ellenbogen, J. M. et al. (2006). The role of sleep in declarative memory consolidation: Passive, permissive, active or none? *Curr Opin Neurobiol 16*(6), 716–22.
5. Guadagni, V. et al. (2014). The effects of sleep deprivation on emotional empathy. *J Sleep Res 23*(6), 657–63.

6. Dahl, R. E. (2007). Sleep and the developing brain. *Sleep 30*(9), 1079–80.
7. Xie, L. et al. (2013). Sleep drives metabolite clearance from the adult brain. *Science 342*(6156), 373–77.
8. Iranzo, A. (2016). Sleep in neurodegenerative diseases. *Sleep Med Clin 11*(1), 1–18; Shamim, S. A. et al. (2019). Insomnia: Risk factor for neurodegenerative diseases. *Cureus 11*(10), e6004.
9. Reutrakul, S., Punjabi, N. M., & Van Cauter, E. (2018, August). Impact of sleep and circadian disturbances on glucose metabolism and type 2 diabetes. In Cowie, C. C., Casagrande, S. S., Menke, A. et al. (eds.). *Diabetes in America* (3rd ed.). National Institute of Diabetes and Digestive and Kidney Diseases.
10. Del Gallo, F. et al. (2014). The reciprocal link between sleep and immune responses. *Arch Ital Biol 152*(2–3), 93–102.
11. Sivertsen, B. et al. (2021). Delayed sleep-wake phase disorder in young adults: Prevalence and correlates from a national survey of Norwegian university students. *Sleep Med 77*, 184–91.
12. Perlow, L. A., & Porter, J. L. (2009). Making time off predictable—and required. *Harv Bus Rev 87*(10), 102–9, 142.
13. Hough, H. (2019). Relationships between Sabbath observance and mental, physical, and spiritual health in clergy. *Pastoral Psychology 68*(2), 171–93; Speedling, B. B. (2019). Celebrating sabbath as a holistic health practice: The transformative power of a sanctuary in time. *J Relig Health 58*(4), 1382–1400; Proeschold-Bell, R. J. (2021). Changes in Sabbath-keeping and mental health over time: Evaluation findings from the Sabbath Living Study. *Journal of Psychology and Theology*.
14. Walker, M. (2017). *Why we sleep*. Scribner.

10

LOVE

While working in the pediatric ICU, I cared for a teenage girl born without a functional left portion of her heart. Jane had survived many complex surgical procedures as an infant, which had allowed her to thrive and flourish. Doing so meant she spent most of her early years in the hospital.

Less than two-thirds of patients diagnosed with hypoplastic left heart syndrome survive at age seven. By the grace of God, Jane had been able to celebrate her fourteenth birthday.

That afternoon, Jane completed a relatively minor surgery compared with previous interventions. Before surgery, I reassured her about the procedure. She was affable and polite. She possessed a gentle spirit and a thankful heart. To her, the healthcare providers were more than names and faces; they were friends and family. She loved to spend time with them while they cared for her needs.

After finishing open heart surgery, the cardiac doctors transferred Jane to the intensive care unit for monitoring. Initially, she was stable, though visibly exhausted.

I examined her surgical site wounds and reviewed pertinent laboratory results. Then I turned my attention to other patients in the ward.

Before I finished rounds, a few of Jane's monitors started beeping. She was only a few hours removed from surgery, yet she seemed increasingly pale. Her capillary refill was sluggish, meaning blood was slow to circulate throughout her body.

We ordered a blood transfusion immediately. But before it arrived, the girl's blood pressure dropped precipitously. Her pulse weakened, and her hands and feet became cold. In this immediate postoperative period, something was terribly wrong. Jane was rapidly losing blood, and signs of cardiogenic shock were ominous.

Without hesitation, the nurses paged the cardiothoracic surgeons. We injected numerous drugs into her body to sustain her heart. Less than fifteen minutes of waiting seemed like hours. The specialists gloved and gowned as they ran to her bedside.

Typically, surgical procedures occur in well-controlled operating rooms, where infection control procedures are meticulously followed. But there was no time to transfer locations. Right in the middle of the pediatric ICU, the heart surgeons opened up Jane's chest. No new incisions were needed—only the freshly tied sutures needed to be cut.

As they performed open-heart surgery in her room, I ordered more blood and prepared medications from the crash cart. It was surreal to see Jane's heart beating inside her chest. The surgeons attempted to stop the bleeding from one of her heart valves while repairing the valve.

While watching this inconceivable event unfold, I couldn't help but think that Jane would not survive. As I shared the news with her mother, tears began streaming down her face. Yet she would not give up hope. Jane's mother believed the Lord had a purpose and a plan for her daughter's life.

Ever since Jane's hypoplastic left heart condition was diagnosed when she was a newborn, her mother recited Jeremiah 29:11 each time her daughter required surgery. Today was no different. "For I know the plans I have for you, declares the LORD," her mother said, "plans for welfare and not for evil, to give you a future and a hope."

After we learned that we shared faith in Christ, we prayed together for Jane.

After many excruciating hours of surgery, dusk turned to dawn. Despite the odds, Jane lived through the night. The initial surgery left her hemorrhaging from the artificial connections created to save her. Miraculously, the cardiothoracic surgeons managed to stop the bleeding while repairing her heart outside the sanitized environment of an operating room.

Later that week, I learned just how challenging Jane's early childhood was and the sacrifices her mother made to provide lifesaving care. From donating blood as a universal donor to sacrificing a career to visit specialists around the country, there was nothing she wouldn't do for her beloved daughter.

Time and time again, God provided healing for this young lady and hope for a mother with unrelenting faith. Through the sacrifice and love of countless people, he mended a broken heart and made it whole once again.

A SYMBOL OF LOVE

For thousands of years, the heart has been linked to love. Greek philosophers believed this vital organ was the source of all emotions. Aristotle even considered it the center of the nervous system and the source of all life.[1] With a primitive understanding of how the heart works—moving and heating blood—it was thought that surely it affected emotions as well.

Ancient Romans thought the left-hand ring finger was connected directly to the heart via the *vena amoris*.[2] The tradition of wearing a wedding ring on this finger began in the Church of England during medieval times. Modern anatomy later disproved the existence of this infamous "vein of love." Yet even today, diamond rings and wedding bands are worn on the left ring finger.

By the 1700s, scientists understood the heart's anatomy better through studying its physiology. After British physician William Harvey developed a circulatory system model, the significance of the heart as the center of emotions began to wane.[3] However, the heart continued to symbolize love in popular culture, poetry, and writing.

Modern medicine considers emotion to be tied primarily to the brain. In particular, a person's limbic system governs their emotional and behavioral responses. The amygdala helps coordinate motivation, emotion, and behavior while reacting to the surrounding environment. Most scientists agree that the hypothalamus plays a crucial role in love, gathering stimuli from our senses and helping create physical responses to associated emotions.

Though the control center of emotion lies mainly in the brain, other organs, like the heart, are intertwined with our feelings. A special connection lies between the heart and the brain.

Broken-heart syndrome, also known as stress-induced cardiomyopathy, is a diagnosable cardiac illness.[4] Patients present with acute chest pain and shortness of breath as if suffering from a heart attack. But tests show no evidence of underlying heart disease. Coronary angiograms reveal arteries without evidence of blockage. Electrocardiograms are devoid of any signs of myocardial infarction. Instead, their pathology is triggered by emotions—broken hearts, commonly experienced after the death of a spouse, betrayal by a partner, divorce, or physical separation from a loved one. Women and the elderly are more likely to succumb to illness after suffering severe emotional distress.

In essence, heartache stuns the heart. This results in limited contraction from the apex of the organ. When severe enough, broken-heart syndrome leads to swelling of the left ventricle, causing the heart to pump blood less effectively.

Unlike patients with heart attacks, those diagnosed with broken-heart syndrome typically recover quickly, sometimes within days or weeks. The heart's function returns to normal as the acute emotional strain lessens.

While physicians no longer define the heart as the source of our emotions, the heart and emotions are still intertwined. The heart is more than just a symbol of love.

LIFE-GIVING BLOOD

Each day, the heart beats approximately one hundred thousand times while pumping more than 1.5 gallons of oxygenated blood through the body every minute. This rich, rubbery substance travels through more than sixty thousand miles of blood vessels. Over an average lifetime, that equates to 2.5 billion heartbeats and a million barrels of blood.

Without this life-giving fluid, the organs of the body cannot survive. They need oxygen from the blood and the nutrients it supplies. Meanwhile, carbon dioxide and waste products are carried within the circulatory system back to the heart, eventually exiting the body.

Though we may consider this dull, red liquid only when bleeding, a drop of blood is teeming with life. It contains more than 5,000,000 red cells, 6,000 to 8,000 white cells, and 150,000 to 450,000 platelets. Water, salts, and proteins make up the remainder of plasma.

When fasting, a person can live weeks without food and a day or two without water. But without oxygenated blood, that same individual will die within minutes. Blood-transported oxygen is essential to cells as they break down sugar for cellular respiration.

For centuries, blood has captured our imagination. Drinking blood was believed to confer supernatural strength and energy in the earliest times. Even today, the nomadic Masai people drink fresh cow's blood mixed with milk. When working in Tanzania, I witnessed this ageless ritual thought to provide healing for the sick.

For Christians, blood symbolizes life and sacrifice. Exodus 12 characterizes the theological significance of this substance during Old Testament times: "The blood shall be a sign for you, on the houses where you are. And when I see the blood, I will pass over you, and no plague will befall you to destroy you, when I strike the land of Egypt" (Ex. 12:13).

During the exodus, Moses led the Jewish people from captivity in Egypt on a journey to the promised land. Initially, nine plagues afflicted Egypt as God prepared to deliver the Israelites from Pharaoh's hand. The Egyptian leader remained stubborn until the tenth and final plague. After witnessing the death of his firstborn son, he was finally convinced to free the Jewish people from slavery.

Passover has become a vital part of the historical redemptive story of Christianity. The night of the tenth plague, God instructed the Israelites to paint their doorposts with blood as they celebrated the evening Passover meal. Modern Egyptologists confirm that these stone doorposts were often inscribed with the names of those living inside the home.

The blood harvested by the Israelites from the Passover lamb marked their homes while also covering their names. Having their Jewish names written on the doorposts was not enough. Being the chosen people of God alone would not save them. Only the blood of offertory lambs could redeem their firstborn children.

According to Leviticus 17:11, blood signified that one life had been sacrificed so another could be saved. The death of a sacrificial animal provided atonement for the sinner. God's covenant with the Israelites was established through the mark of blood.

What was initiated by the Lord for Jews was completed by the death of Jesus Christ for all people. In 1 Corinthians 11:25–26, Jesus says, "This cup is the new covenant in my blood. Do this, as often as you drink it, in remembrance of me. For as often as you eat this bread and drink the cup, you proclaim the Lord's death until he comes."

The Messiah heralded by John the Baptist became the

Lamb of God. He accomplished what an infinite number of animal sacrifices could not, taking away the world's sins (John 1:29). At Calvary, Jesus negated the obligation of priests and animal sacrifices. High priests ascended the altar to sacrifice animals with their bare hands. Yet in perfect love, the Father lifted his hands from Jesus so that the sin of the world would descend on him.

Leviticus confirms that the life of the flesh is found in the blood (17:11). Without blood, physical life is untenable. Vital organs malfunction. Temperature becomes dysregulated. Tissues die. The body succumbs to shock.

So it is no coincidence that God associates the substance of blood with eternal life. Through the incarnation of Christ, he put on flesh and bone while receiving a heart to circulate blood. Yet he willingly gave up this life-sustaining liquid so others could live beyond their earthen vessels.

The juxtaposition of an infinite number of animal sacrifices with the willing submission of the Savior alone demonstrates the beauty of the gospel. Through this metaphoric transfusion of Jesus' blood, all inherit the life-giving effects of his sacrifice.

We cannot survive without blood. Yet we cannot have abundant life without consuming the blood of Jesus. In the past, God prohibited Noah from drinking the blood of animals (Gen. 9:4). But with the incarnation and resurrection of Jesus, Christ invites us to "drink" his blood and inherit his life.

Only the blood of Christ gives us eternal life while allowing us to love sacrificially. For whoever "feeds on my flesh and drinks my blood abides in me, and I in him" (John 6:56).

CREATED TO LOVE

In Genesis 1, God creates humanity to rule over birds soaring through the air, fish swimming in the seas, and animals roaming dry land. In man, he crafts a being in his image and likeness. Realizing it is not good for the man to be alone, God fashions a woman as his companion. Together, they enjoy fellowship with one another.

Adam and Eve were created to love. In the garden of Eden, they experienced perfect love—between each other and with God. This love was pure and untainted, healthy and fulfilling.

But after eating fruit from the forbidden tree, Adam and Eve were cast out of the garden. A once-perfect love became fractured and broken. Genesis 3:16 confirms the enmity between man and woman. All future descendants would need to learn how to love and be loved. They would receive ample opportunities to do so. In time, Jesus modeled and taught us how to love through his incarnation. First John 4:7–8 encourages us to "love one another, for love is from God, and whoever loves has been born of God and knows God. Anyone who does not love does not know God, because God is love."

Science confirms what our faith asserts, that we cannot live without love. The widowhood effect demonstrates the consequence of losing a loved one.[5] When a spouse dies, the recent widow or widower is more likely to suffer the same fate.

The risk of death is between 30 percent and 90 percent more among widows in the first three months after a spouse dies. After that, it plateaus at 15 percent.[6] No matter what the cause of death is, the widowhood effect persists among men and women of all ages. Some causes of death have a more significant impact than others; not surprisingly, this influence is magnified when partners die suddenly.

Unmarried adults are also more likely to die younger than their married counterparts.[7] Though this is true for all categories of available individuals, it is most striking for those who never marry.[8] In addition, unmarried men have a higher chance of dying than their female counterparts.

Accumulated research suggests that social isolation is a significant cause of premature death. Loneliness and social isolation can increase the likelihood of death by 26 percent to 32 percent.[9] Presumably, marriage is an indicator of social connectedness while protecting against loneliness. Being married invites spouses to share in the fullness of love.

In his classic book *The Four Loves*, C. S. Lewis describes the meaning of four Greek words that signify love. These words have become synonymous with the way Christians understand biblical love. Even those without formal language training can identify several of these terms.

In the Greek language, more than four words signify love. Lewis chose to focus on *agape*, *phileo*, *eros*, and *storge*, given their direct references in the New Testament or value in defining biblical love. Other Greek words depict mature (*pragma*), playful (*ludus*), and obsessive (*mania*) love, in addition to self-love (*philautia*).

The most famous love is *agape*, which Lewis described as charity. *Agape* is perfect, unconditional love. There are no strings attached and no expectations in return. According to 1 John 4:8, God's love is *agape*. His devotion to all humanity is demonstrated through the sacrifice of his only Son.

In the same way, Jesus demonstrated unconditional

love for the world by willingly dying on the cross. There, he redeemed our sins once and for all on Good Friday. This sacrificial act exemplifies charity without consideration for oneself.

First Corinthians 13 defines *agape* as patient and kind, without envy or boasting, not arrogant or rude. Unconditional love does not insist on its own way. It is not irritable or resentful, never rejoicing at wrongdoing, but rejoicing only with the truth. *Agape* bears, believes, hopes, and endures all things.

During his earthly ministry, Jesus instructed his disciples to demonstrate this selfless love to one another. They were to love brothers and sisters in the same way he modeled *agape*. Through their acts of charity, nonbelievers would know they were his followers (John 13:35).

The Greek word *phileo* frequently appears in the New Testament. The city of brotherly love, Philadelphia, is named after this type of love. *Phileo* is found in deep and meaningful friendships. Close friends share common passions and likeminded thinking that naturally leads to their love for one another. *Phileo* is the backbone of their relationship as kindred spirits.

In Romans 12:10, Paul uses *phileo* to describe the way Christians are to be devoted to one another. Love for neighbors and friends should be similar to that reserved for siblings. Ancient Greeks and Romans valued *phileo* greatly; unfortunately, it has become deemphasized in modern times.

The word *storge* is the love prevalent among families. Lewis describes it as affection. Natural bonds occur through blood or marriage. *Storge* expresses the characteristic love of a parent for their children, a grandparent for their grandchildren, and between siblings. It occurs instinctively among those closest to us.

In the New Testament, *storge* appears in its compound form as Paul exhorts Christians to love one another with brotherly affection. The negative form illustrates a lack of affection of the unrighteous (Rom. 1:31) and the disappearance of natural love among family members during the end times (2 Tim. 3:3).

Finally, *eros* is the root of the English word *erotic*. Erotic love is sexual desire for another person. In Greek mythology, Eros shoots arrows into people's hearts, causing them to fall madly in love with one another. Lewis cautioned against this unbridled passion as one of the most dangerous ways to lose control and commit sin.

Though *eros* never appears in the New Testament, its presence is scattered throughout the Word. Song of Solomon describes in explicit detail the erotic joy between a newlywed bride and groom. In 2 Samuel, David cannot escape the slippery slope of lust when he sees Bathsheba bathing on her rooftop. The flames of desire send him to his downfall, as he eventually commits adultery with her and then commits murder to cover up his sin.

Whether it's *agape*, *phileo*, *eros*, or *storge*, we are created to love. Experiencing and reciprocating love in the ways God ordained, we move one step closer to the garden of Eden. Though original sin ruined the initial opportunity for perfect love, the gift of the Holy Spirit renewed the prospect of loving the way we were designed to.

As temples of the Holy Spirit, our bodies have the capacity not only to receive love but also to give it away freely. The Spirit convicts us to love even when our hearts are hardened. He tempers unbridled emotions that threaten to poison healthy love. He invites us to love our neighbors as ourselves.

Unconditional love is antithetical to our sinful nature. To love with no expectations is rare. Yet we are called to do so sacrificially as Christ did. Only the Holy Spirit can empower us to act in such ways, rising beyond our flesh to love beyond imagination.

Because the Holy Spirit dwells in us, we can love incarnationally.

DISTORTED LOVE

Preparing for college, an eighteen-year-old high-school senior once visited me for her well-child check and to ensure she had the required immunizations to start school. As I reviewed Mary's record, there were no apparent concerns. She did not suffer from any chronic health problems and took no medications.

I had not seen Mary since she missed her seventeen-year-old annual exam. Her medical chart was sparse, except for yearly physicals and an occasional sick visit. Tall and thin, she consistently had a low body-mass index.

As I walked into the exam room, Mary looked at her face in the mirror. It was striking to see how much she had changed in a couple of years. Before, she never wore makeup and typically sported comfortable clothing. Today, her outfit was stylish and fashionable. Eyeliner and lipstick hid the innocence of youth.

I asked whether she had any concerns about her skin. It's common for teenagers to deal with acne and request treatment. Instead, Mary commented about her nose being imperfect and how she didn't like her puffy lips. I assured her that she had lovely facial features. She replied that the other girls in her modeling agency were picture perfect. Unlike her, they were born without flaws.

Ever since Mary started modeling, her life and demeanor had changed. No longer was she a free-spirited teenager who loved to laugh and smile. Now she obsessed over her looks and worried about perceived imperfections in her body. Mary spent much time comparing herself with other girls. Her obsession was affecting her schoolwork, family, and relationships.

These symptoms are indicative of body dysmorphic disorder. In this condition, a person can't stop thinking about their body's flaws. Even minor imperfections become significant distractions. Patients get stuck on their appearance, believing their flaws make them undesirable to others. They repeatedly check the mirror throughout the day.

Those with body dysmorphic disorder often seek reassurance from others. They may try to fix their perceived flaws. Usually, this leads to a cycle of cosmetic procedures and medical interventions. Over time, the repetitive behaviors and fixation wear on a person's ability to find joy while loving others.

Several treatments exist for body dysmorphic disorder. Cognitive behavioral therapy can help with negative thoughts, emotions, and behaviors. Serotonin reuptake inhibitors target chemical imbalances.

But at the core, the pathology stems from a lack of self-love. Patients with low self-esteem do not fully comprehend God's love for them. Each detail, every curvature, and all of their features were carefully crafted to make them unique in his eyes. He created their bodies perfectly in his image.

If they experienced the fullness of God's love, they would know how precious they are to him. Their bodies are fearfully and wonderfully made (Ps. 139:14). Even the hairs on their head are numbered (Luke 12:7). In some cases, individuals experience irrational hatred of their bodies secondary to a chemical imbalance or latent pathology. For these people, medical interventions may be necessary.

Though created for good, love can become distorted. Lewis argues that natural loves (*phileo*, *storge*, *eros*) must be in proper balance with God's *agape* love. The former become perverted when our focus on them is disproportionate to our love for God.

While this is true, we must understand and experience God's unconditional love to foster the right balance of natural love. Without properly embracing God's love for us, we risk filling the gap with other loves. The spiritual disciplines that draw us closer to the love of God prevent us from being pulled into unhealthy forms of *phileo*, *storge*, and *eros*.

When we are brought to repentance by Jesus' sacrifice, we know the love of Christ. Our hearts are filled with joy as our voices reach to the heavens in a symphony of praise. If we submit to the still, small voice of the Holy Spirit, our souls feel the heartbeat of God. The more we experience the love of God, the less we depend on other forms of love.

While we were created to receive the richness of God's love, our bodies were also designed to experience natural love in life-giving ways. When *phileo*, *storge*, and *eros* are suppressed, these loves may erupt in harmful behaviors. In contrast, if any of these loves are left unchecked, they tend to carve out a more significant proportion of our hearts over time.

Storge is instinctive for families. Newborn infants

mesmerize their parents in the first year of life. Grandparents are known for spoiling their new kin while demonstrating the fondest of affection. Little effort is required to focus on our families, sometimes to the detriment of others. But when their health, well-being, and livelihood become the center of our lives, our affection becomes distorted. C. S. Lewis's *The Great Divorce* describes a poignant fictional example of this, as a mother's distorted love for a son prevents her from experiencing the grace of God and the joy of heaven.

Jesus knew how subtle this shift could be, utterly unnoticeable at times. That's why he uses the strong word *hate* to describe how we should treat our mothers and fathers, brothers and sisters if we want to be his disciples (Luke 14:26). Christ does not demand that we take the word at face value. Instead, he uses hyperbole to illustrate how easily *storge* can become an idol. The deep cost of discipleship demands that we love our families with a proper balance.

Life-giving relationships result from the bonds of *phileo*. The relationship of kindred spirits is filled with this love. Groomsmen and bridesmaids effortlessly withstand the challenges of distance and time as if they were never separated. Laughter fills the air when they reminisce on the past, and these memories strengthen their mutual love.

But when our friendships become exclusive, they drift away from the intent of brotherly love. *Phileo* was never meant to be self-centered. Cliques and divisive self-interest groups destroy the testimony of the church. But the powerful bonds forged by *phileo* contribute to meaningful relationships while forming solid communities. What a blessing when they unite Christians to engage in the kingdom's work.

Finally, *eros* is a gift of God that encourages spouses to

enjoy his creation while becoming one flesh. The Creator designed our bodies to fulfill the longings of the flesh while simultaneously honoring God. Sexuality and sensual desire were always meant to be a blessing.

But when we circumvent God's plans, *eros* becomes perverted. Though adultery has plagued humanity since its inception, the challenges of sexual sin are more pronounced than ever. Once hidden behind newspaper stands, pornography is now easily accessible through the privacy of home computers and smartphones. Up to 70 percent of adults ages eighteen to thirty admit to being tempted by it. The Scriptures warn about the slippery slope of sexual temptation. More than fifty times, the topic of infidelity is addressed, including its mention in the Ten Commandments.

First Peter 5:8 reminds us that we must be vigilant and on guard, for the devil prowls like a roaring lion waiting to devour us. Whether it's *eros*, *phileo*, or *storge*, we must keep every love in check. Any one can become distorted and destructive. Yet every one was created to be uncorrupting and life-giving.

LOVE THE LORD, THEN YOUR NEIGHBOR

Mark 12 quotes the two greatest commandments of Scripture: "'You shall love the Lord your God with all your heart and with all your soul and with all your mind and with all your strength.' . . . 'You shall love your neighbor as yourself'" (Mark 12:30–31). We cannot live without love. Our hearts yearn to experience it. They harden, crack, and eventually break without it. First John 4:19 says, "We love because he first loved us." Receiving God's unconditional love prepares us to reciprocate it while loving others.

The imperative to love God with all our hearts, minds, souls, and strength speaks to the centrality of our relationship with him. Framing love in this context makes it the critical motivation for every decision. Love for God strengthens our resolve, inspires our thinking, captivates our hearts, and awakens our souls.

These words from Mark 12 describe the individual means to love the Lord, but they also represent a collective response to God's love for us. Our devotion is not to be fragmented and disjointed but all-encompassing and synergistic. Every glance, taste, smell, heartbeat, breath, and movement is for the Lord. Together they form a symphony of adoration that echoes throughout our lives.

Only by experiencing and reciprocating God's love can we love others. Since it is the foundation of our love, the first commandment precedes loving our neighbors. Once we love God with our entire being, we can love those around us. When we neglect the former, our love for others becomes distorted.

Who is our neighbor? Jesus answers the Pharisee's question with the parable of the good Samaritan (Luke 10:25–37). A Jewish man traveling from Jerusalem to Jericho is beaten, stripped, and left for dead by a pack of robbers. Both a priest and a Levite pass by the wounded man only to leave him alone to die. But the Samaritan does not cast a blind eye. Samaritans were considered outcasts by the Jews, since they were part Jew and part gentile.

The person least expected to show compassion to the man pours oil and wine on his wounds and bandages them.

Then he places him on his donkey and brings him to a nearby inn. The next day he pays the innkeeper two denarii to nurse the man back to health. Before leaving, he promises to pay any additional expenses incurred.

After sharing the story, Jesus asks the Pharisee to identify the one who was a good neighbor. He replies, "The one who showed him mercy." To which Jesus says, "You go, and do likewise" (v. 37). By casting the Samaritan as the hero, Jesus teaches us how to show compassion to those we despise. He demonstrates what embodied love for those who hate us looks like.

Loving our neighbor is intuitive when we are kindred spirits. The bonds of *phileo* love require little sacrifice. But what about a complete stranger or someone socially, culturally, or ethnically different from us? Our neighbors are those who are both near and far away. When God orchestrates opportunities to love others, we must act. It matters not who they are but whether we are willing to love them.

"So now faith, hope, and love abide, these three; but the greatest of these is love" (1 Cor. 13:13).

DEATH, RESURRECTION, AND THE RISEN KING

Again, as 1 John 4:19 says, "We love because he first loved us." Each of us can love others because of the Word made flesh, Jesus. Jesus' sacrifice allows us to live and to love others incarnationally through the power of the Spirit.

When our health fails and medical prospects are bleak, Christ alone gives eternal hope. Ultimately, the wages of sin is death. No human being can circumvent the inevitable.

Even for those born with healthy bodies, the process of aging occurs rapidly. With time, our bodies lose control of our cells and genes. Every nerve, tissue, and organ eventually stops working.

Only in eternity will illness and disease cease to exist and our bodies be made whole. Yet incarnational health choices are a reflection of the resurrection today. The risen King gives meaning to our present lives, both physically and spiritually. Through him, we are able to glorify God and point others to the potential for future resurrection.

How glorious will our heavenly bodies be on that day. No longer will those vessels be susceptible to illness, disease, and aging. Instead, they will live impervious to the effects of sin and decay. Paul describes these ethereal bodies in 1 Corinthians 15:42–44: "What is sown is perishable; what is raised is imperishable. It is sown in dishonor; it is raised in glory. It is sown in weakness; it is raised in power. It is sown a natural body; it is raised a spiritual body. If there is a natural body, there is also a spiritual body."

On that day, our bodies will be raised in power. Inherited genetic diseases and chronic debilitating illnesses will not ravage our bodies. Think of a day when your muscles no longer ache and your bones cannot break. Heart attacks will not steal life away. Alzheimer's disease cannot erase the memories of loved ones.

In the fullness of time, we will be sanctified wholly. Our souls will be in harmony with the Holy Spirit. No longer will we be subjected to the weaknesses of the flesh. Instead, our resurrected bodies will work in perfect concert with the spiritual desire to worship God.

On that day, "he will wipe away every tear from their eyes, and death shall be no more, neither shall there be mourning, nor crying, nor pain anymore, for the former things have passed away" (Rev. 21:4).

In that moment, we will echo the words of Paul, saying, "O death, where is your victory? O death, where is your sting?" (1 Cor. 15:55–57). Ultimately, the Great Physician triumphs over all that afflicts our spiritual and physical health. It is possible only through the love of a Father by the sacrifice of his Son.

QUESTIONS FOR REFLECTION

1. Modern medicine considers the brain to be the seat of emotions. But illnesses like broken-heart syndrome demonstrate a connection between the heart and mind. Have you or a loved one experienced this juxtaposition?
2. The substance of blood is life giving, for both our physical bodies and our spiritual ones. Explain the significance of blood within our earthly lives and for our future inheritance.
3. What are four major types of love described in the Bible? How is the Holy Spirit convicting you to demonstrate each?
4. First John 4:7–8 encourages us to "love one another, for love is from God, and whoever loves has been born of God and knows God. Anyone who does not love does not know God, because God is love." We were created to love. Yet sometimes we distort our love. How have you engaged in unhealthy love, and what were the repercussions?
5. What is the greatest commandment, and how can you embrace it?

6. How does the future resurrection of our bodies affect our physical and spiritual health today?

NOTES

1. Clarke, E. (1963). Aristotelian concepts of the form and function of the brain. *Bull Hist Med 37*, 1–14.
2. Kunz, G. F. (1917). *Rings for the finger: From the earliest known times to the present.*
3. Harvey, W. (1998). *Exercitatio anatomica de motu cordis et sanguinis in animalibus.* Octavo.
4. Sharkey, S. W. et al. (2005). Acute and reversible cardiomyopathy provoked by stress in women from the United States. *Circulation 111*(4), 472–79; Wittstein, I. S., Thiemann, D. R., Lima, J. A. C., Baughman, K. L., Schulman, S. P., Gerstenblith, G., Wu, K. C., Rade, J. J., Bivalacqua, T. J., Champion, H. C. (2005, February 10). Neurohumoral features of myocardial stunning due to sudden emotional stress. *N Engl J Med 352*(6), 539–48. doi: 10.1056/NEJMoa043046
5. Schaefer, C. et al. (1995). Mortality following conjugal bereavement and the effects of a shared environment. *Am J Epidemiol 141*(12), 1142–52; Elwert, F., & Christakis, N. A. (2008). The effect of widowhood on mortality by the causes of death of both spouses. *Am J Public Health 98*(11), 2092–98.
6. Schaefer, C. et al. (1995). Mortality following conjugal bereavement and the effects of a shared environment. *Am J Epidemiol 141*(12), 1142–52; Martikainen, P., & Valkonen, T. (1996). Mortality after the death of a spouse: Rates and causes of death in a large Finnish cohort. *Am J Public Health 86*(8), 1087–93.

7. Lahorgue, Z. (1960). Morbidity and marital status. *J Chronic Dis* 12, 476–98; Verbrugge, L. M., & Wingard, D. L. (1987). Sex differentials in health and mortality. *Women Health* 12(2), 103–45.
8. Kaplan, R. M., & Kronick, R. G. (2006). Marital status and longevity in the United States population. *J Epidemiol Community Health* 60(9), 760–65.
9. Holt-Lunstad, J. et al. (2015). Loneliness and social isolation as risk factors for mortality: A meta-analytic review. *Perspect Psychol Sci* 10(2), 227–37.